FAMOUS REGIMENTS

The Wiltshire Regiment

Other titles in this series

THE ROYAL FUSILIERS
THE ROYAL NORFOLK REGIMENT
THE QUEEN'S ROYAL REGIMENT (West Surrey)
THE KING'S ROYAL RIFLE CORPS
THE SOUTH WALES BORDERERS
THE GREEN HOWARDS
THE ROYAL BERKSHIRE REGIMENT
THE SCOTS GUARDS
THE HIGHLAND LIGHT INFANTRY
THE 17TH/21ST LANCERS
THE 11TH HUSSARS
THE BLACK WATCH
THE ROYAL HAMPSHIRE REGIMENT
THE KING'S OWN YORKSHIRE LIGHT INFANTRY
THE GORDON HIGHLANDERS
THE YORK AND LANCASTER REGIMENT
THE ARGYLL AND SUTHERLAND HIGHLANDERS
THE WILTSHIRE REGIMENT
THE SOMERSET LIGHT INFANTRY

FAMOUS REGIMENTS

Edited by
Lt.-General Sir Brian Horrocks

The Wiltshire Regiment

(The 62nd and 99th Regiments of Foot)

by Tom Gibson

Leo Cooper Ltd., London

First published in Great Britain, 1969
by Leo Cooper Ltd.,
47 Museum Street,
London W.C.1

85052 007 X

Printed in Great Britain
by Ebenezer Baylis and Son, Ltd.
The Trinity Press, Worcester, and London

INTRODUCTION TO THE SERIES

by Lt.-General Sir Brian Horrocks

IT IS ALWAYS sad when old friends depart. In the last few years many famous old regiments have disappeared, merged into larger formations.

I suppose this is inevitable; strategy and tactics are always changing, forcing the structure of the Army to change too. But the memories of the past still linger in minds now trained to great technical proficiency and surrounded by sophisticated equipment. Nevertheless the disappearance of these well-known names as separate units marks the end of a military epoch; but we must never forget that, throughout the years, each of these regiments has carved for itself a special niche in British history. The qualities of the British character, both good and bad, which helped England to her important position in the world can be seen at work in the regiments of the old Army. To see why these regiments succeeded under Marlborough and Wellington yet failed in the American War of Independence should help us in assessing the past.

Though many Battle Honours were won during historic campaigns, the greatest contribution which our Regiments have made to the British Empire is rarely mentioned: this has surely been the protection they have afforded to those indomitable British merchants, who in search of fresh markets spread our influence all over the world. For some of these this involved spending many years in stinking garrisons overseas where their casualties from disease were often far greater than those suffered on active service.

The main strength of our military system has always lain in the fact that regimental roots were planted deep into the

British countryside in the shape of the Territorial Army whose battalions are also subject to the cold winds of change. This ensured the closest possible link between civilian and military worlds, and built up a unique county and family *esprit de corps* which exists in no other Army in the world. A Cockney regiment, a West Country regiment and a Highland regiment differed from each other greatly, though they fought side by side in scores of battles. In spite of miserable conditions and savage discipline, a man often felt he belonged within the regiment—he shared the background and the hopes of his fellows. That was a great comfort for a soldier. Many times, at Old Comrades' gatherings, some old soldier has come up to me and said, referring to one of the World Wars, 'They were good times, Sir, weren't they?'

They were not good times at all. They were horrible times; but what these men remember and now miss was the comradeship and *esprit de corps* of the old regular regiments. These regiments, which bound men together and helped them through the pain and fear of war, deserve to be recalled.

Regimental histories are usually terribly dull, as the authors are forced to record the smallest operation and include as many names as possible. In this series we have something new. Freed from the tyranny of minute detail, the authors have sought to capture that subtle quarry, the regimental spirit. The history of each regiment is a story of a type of British life now fading away. These stories illuminate the past, and should help us to think more clearly about the military future.

The Wiltshire Regiment

A SPECIAL INTRODUCTION

by Lt.-General Sir Brian Horrocks

IN 1949 when I was invalided out of the Army and appointed Gentleman Usher of The Black Rod in the House of Lords we went to live in a small hamlet called Chisbury, on the edge of Savernake Forest in Wiltshire. In our nearest village, Great Bedwyn, there existed a most flourishing branch of the British Legion, which I automatically joined.

Each summer the Legion held a special parade on what was called 'Comrades' Sunday'. This was the most important day in the village calendar—the girls put on their new frocks and detachments attended from all the neighbouring branches of the Legion.

In charge of the parade was a young R.S.M., selected solely because of his powerful word of command, because quite rightly there was no rank in the Bedwyn Legion—we were all equal.

As my last appointment in the Army had been Commander-in-Chief British Army of the Rhine, and I had not really marched in the ranks for a good many years, I approached my first parade with a certain amount of trepidation.

'1914-18 War on the right—last war on the left,' bellowed the R.S.M. from a commanding position on a bank at the side of the road. As I had been in both wars I positioned myself in the centre.

'Right Turn—Quick March', and off we went to the strains of the Kennett Silver Band which had been hired for the day. I went swinging along and began to congratulate myself that I was not really doing too badly. Then suddenly I heard a voice behind me saying in a good Wiltshire accent—'I wish that old bastard in the green hat would keep covered off'.

He was, I am afraid, alluding to me and I realised once and for all that the war was really over.

The War Memorial was by the village church, which was of considerable historic interest, as Henry VIII is reputed to have worshipped here when he was courting Jane Seymour, who lived nearby.

As I stood there listening to Reveille and the Last Post surrounded by the beautiful Wiltshire countryside, I looked along the weather-beaten faces of the men on either side of me, many of whom tended this land, and I felt in a curious way closer to the heart of this fine old country of ours than I have ever done before or since.

I knew also that on this parade were men who were the third and possibly fourth generation of the same family which had served in the County Regiment and whose ancestors might well have been in the 62nd Foot (subsequently renamed 1st Battalion Wiltshire Regiment) formed in Exeter in 1756, or in the 99th which originated in Glasgow in 1824 but became the 2nd Battalion Wiltshire Regiment under the Cardwell system, by which battalions were linked together and based on particular counties or large cities. An odd marriage this—between Southerner and Lowlander, but no alliance could have been happier.

It was a moving moment, for me, because I felt a deep sense of gratitude to these West Country men. After all it was their 4th Territorial Battalion alongside the 5th Somerset Light Infantry from the neighbouring county, both led by their commanding officers in person, Lieutenant-Colonels Luce and Lipscomb, and preceded by two troops of the 13th/18th Hussars Tanks, which had occupied Mount Pincon. This was the dominating feature of all this part of the countryside and they had thus opened the door for the breakout from the Normandy beach head. It was also the 4th Somerset Light Infantry and 5th Wilts which had seized the bridgehead over the river Seine, in one of the most difficult and confused river crossings of the last war. This enabled me to pass through some 600 tanks for the 300 mile thrust through the heart of the enemy to Brussels, which we liberated six days later.

No wonder I have always had a deep admiration and affection for these West Country Units in the 43rd Wessex Division with its famous Wyvern badge.

The main difficulty which faces all the authors in the 'Famous Regiments' series is how to deal adequately, in such a short history, with the vast canvas of the 1914-18 War. This is particularly so in the case of the Wiltshire Regiment, because their battalions served in almost every theatre of war—Egypt, Gallipoli, Mesopotamia, Palestine, Greece, Salonika, and of course on the Western Front where time after time, like so many others they were practically wiped out. On one occasion all that remained of the 6th Battalion was one officer, one sergeant, eighteen other ranks and the officer was wounded.

Yet Lieutenant-Colonel Tom Gibson, the author, has succeeded admirably in producing a coherent story in which the spirit of the Regiment breathes from every page. It is unquestionably the individual stories which he has so painstakingly unearthed which account for this atmosphere. I have never read an account of a more gallant action than when Captain R. J. F. Hayward, already an M.C., who started in the 6th Battalion, won the V.C. when serving with the 1st.

But for me the spirit of the Regiment is personified in Brewer, who joined as a band boy in 1905 and ended up forty-four years later as a much respected Major and Quartermaster. Moreover his personal account of an attack on the Western Front as experienced by an infantry N.C.O. is a most vivid piece of writing.

The country owes a great debt to the 'Brewers' of the old Regular Army, who are the salt of the earth.

I congratulate Tom Gibson on producing a book which anyone connected with Wiltshire or the Wiltshire Regiment will not only find fascinating reading, but of which he should also be justly proud.

Acknowledgements

The author and publishers wish to express their gratitude to Lieutenant-Colonel E. A. T. Boggis for his very ready help regarding the provision of illustrations from the Regimental Museum, Le Marchant Barracks, Devizes, and his wholehearted co-operation in general. The author would also like to thank Lieutenant-Colonel R. J. F. Hayward V.C., M.C., for his prompt and courteous assistance in identifying the other officers in the group photograph on pp. 115

Chapter I

The Seeds of Empire

In October, 1756, an unwonted military bustle stirred the sleepy cathedral town of Exeter. Drafts of soldiers arrived, and through the narrow, cobbled streets red-coated recruiting parties moved about the taverns and market places. Along winding roads that threaded through the rolling hills and rich red-soil fields of Devon, these persistent recruiters, with beat of drum, toured out to the other towns and hamlets of the county. Britain, with Prussia as an ally, was at war with France and Austria, and the formidable power of France now caused some frantic strengthening of the Royal Navy and the Army. It was decided in Whitehall, and the Army Order was issued from the Horse Guards on September 20, 1756, that the fifteen senior Regiments of Foot would have second battalions. The new battalion of the 4th Foot was to be raised at Exeter.

Who were those nameless recruits, now long lost in the dust of history, who gathered at Exeter in that first autumn of the Seven Years War, to be drilled and cowed into some semblance of order by heavy-handed and short-tempered veteran N.C.O.s? Like the rank and file of all eighteenth-century units, they were the compulsory transfers of the unwanted from other Regiments, gawky country yokels and the half-starved sweepings of the city streets (any Parish Officer could impress a vagrant, a lucrative sideline that gained him £1 and the Parish the other £2). Recruiting standards had a minimum height of five feet four inches (down to five feet two inches after four years of war) for an engagement of five years; the recruit, if a volunteer, gained a bounty of £3 which newly-met old soldier mates in his first billet promptly proceeded to drink. Discipline was maintained by liberal use of the lash, food was rudimentary and generally appalling, and life was only made bearable by soporific bouts of drinking which took the few remaining pence of daily pay left from all the mulctary stoppages.

The lash and harsh conditions of the British soldier were to exist for yet another century, but these dregs of a nation wholly bent on commerce and gain, when brought together in a regiment where they were cared for by sound, professional officers, bore great privations with an incredibly stoic cheerfulness. Such men as these embryo reinforcements for the 4th Foot, under inspiring leadership, were to carve out an empire for their parsimonious homeland during the Seven Years War in the great forests of Canada and in the terrible heat and savage splendour of far-away India.

But glory in distant fields was far removed from the bewildered minds of the raw recruits at Exeter that October. At the end of the month, however, came a respite from the constant drill: the parent regiment, the 4th Foot, disembarked at Torbay after serving in the defeated garrison of Minorca and its infant second battalion marched south along the River Exe and its estuary and over the wooded hills to join it.

In May, 1757, the 4th Foot moved to Plymouth and later, that October, its new 2nd Battalion was ordered to be prepared to do duty with the Fleet as marines. A few months later, in January, 1758, four companies embarked in H.M.S. *Bedford*, *Nottingham*, *Prince of Orange*, *Burford* and *Centurion*, ships of Admiral Boscawen's fleet soon to sail for Halifax, Nova Scotia, the mounting base for a seaborne attack on French Canada.

However, on April 21, when these four companies were with the fleet on passage across the Atlantic and Battalion Headquarters and the remaining companies served on at

Plymouth, a drastic re-incarnation of 2/4th Foot occurred. The Horse Guards now decreed that the fifteen new second battalions throughout the Army would be re-numbered as separate Regiments, from 61 to 75, and the 2/4th Foot, under this reorganization, became the 62nd Regiment. A modern historian of the Regiment, Colonel N. C. E. Kenrick, D.S.O., gives a sartorial picture of the new Regiment about this time in his *The Story of the Wiltshire Regiment*:

> 'The 62nd's uniform consisted of a scarlet coat with yellowish buff facings; the waistcoat, breeches and coat-linings were also yellowish buff. They wore white gaiters except on active service, when they were changed to black. The Regimental lace was silver with two blue and one straw-coloured stripe. The blue marked the association with the 4th Foot, a Royal Regiment; the straw colour and the yellowish buff also came from the 4th. In 1684 it had been known as the 2nd Tangier Regiment, and was created the "Duchess of York and Albany's" by Charles II, who gave the Regiment yellow facings, the Duke of York's favourite colour. As the 62nd could no longer wear the Royal blue facings on ceasing to be the 2/4th, they probably chose yellowish buff from their previous associations. Hair was powdered and pig-tailed. All ranks wore hats turned up on three sides. There was little distinction by ranks. Officers wearing a red sash over the right shoulder, and sergeants a red sash, usually with regimental facings, round the waist.'

One month after the re-titlement, the 62nd Regiment was blooded in action for the first time.

On May 27, 1758, the French outposts overlooking Gabarus Bay, on Cape Breton Island, stood to arms as they beheld a formidable spectacle. Thirty-nine British warships sailed into the Bay and anchored, followed over several days by a great assembly of transports crammed with 11,000 troops. By June 2, the concentration of the invading force under the command of Admiral Boscawen, and with 400 men of the 62nd on board five of his men-o'-war, was

Wiltshire Newspapers Ltd

Louisbourg, 1758, from Lighthouse Point, showing the town and fortress, with the French men-o'-war in the harbour, and the British fleet in Gabarus Bay in the distance.

complete. The siege of Louisbourg, 'the Gibraltar of Canada', was about to be mounted.

Louisbourg, which had been built at great expense by the French over a period from 1713 to guard the lifeline of communications between the Gulf of St. Lawrence and France and also, not least, the rich cod fisheries in the local waters from the powerful and expansionist New Englanders, was sited on a small promontory in its own spacious harbour. Its garrison of 3,000 regulars, 2,600 seamen and about 400 Canadian militia, together with the five ships of the line that lay in harbour, was under the overall command of Governor Drucour.

After an opposed assault landing in Freshwater Cove, ably commanded by the thrusting Brigadier James Wolfe, the labour to establish the expedition ashore and deployed in siege lines was immense. While these preparations went on, Major-General Amherst ordered Wolfe to take 1,200 men and artillery to the abandoned French battery position at Lighthouse Point, on the far side of the harbour mouth, to engage the enemy coastal battery on Goat Island, which commanded the harbour entrance, and also the five enemy warships lying at anchor. Soon, the island battery was silenced and the ships driven in to the close protection of the fortress.

The French reacted vigorously to this prising-open of the harbour entrance. One foggy night soon afterwards, they sank six blockships in the channel and began a series of attacks on the Lighthouse Point gun positions. The 62nd and other marines, with heavy guns from the Fleet, had now relieved Wolfe and his force, to allow their return to the main siege lines, and men of the 62nd, dug in to give infantry protection to the guns, saw brisk action in repelling the desperate French assaults.

In July, Amherst began his operation in earnest, pushing trenches forward towards the Dauphin Bastion in the north and the Princess Bastion in the south. The British bombardment thudded out and crashed down on Louisbourg day after day, gradually reducing the proud fortress, and the siege drew inexorably to the inevitable end. On July 21, the naval battery at Lighthouse Point, still defended by 62nd detachments, scored an excellent hit on one of the French ships of the line; blazing furiously, it drifted to collide with and set alight two other ships. Some nights later, 600 British seamen and marines, including the men of the 62nd not at Lighthouse Point, notably from H.M.S. *Burford* and *Bedford*, rowed stealthily into Louisbourg harbour and captured the two remaining warships. One went aground and had to be burned, the other was towed away.

This cool and audacious exploit seems to have been the snapping point for the tired and hard pressed French defence. Moreover, a practicable breach had been opened, inviting a general assault by the British. On July 26, Governor Drucour surrendered to Boscawen unconditionally. Some 5,500 French soldiers and seamen were taken prisoner, with over 200 pieces of artillery and large stocks of ammunition and stores, from a fortress and naval base whose strategic concept was doomed once the French had lost local command of the sea.

Amherst sailed away to Boston with the 1st, 17th, 47th, 48th Foot and Fraser's Highlanders, leaving a garrison at Louisbourg of the 22nd, 28th, 40th and 45th Foot. He also left a final mission for James Wolfe, in concert with Rear-Admiral Sir Charles Hardy's squadron, to devastate the Gaspé coast on the southern shoulder of the Gulf of St. Lawrence. The 62nd took part in this unpleasant operation, and was to see yet more of the brilliant Wolfe during the following year in the campaign against Quebec.

When the mastheads of the first British sail appeared in the St. Lawrence before the astonished garrison and citizens of Quebec on June 21, 1759, the consternation in the French command was considerable. A British invasion of Canada had been confidently expected from Lake Ontario in the west and from the traditional route of Lake Champlain in the south but, even after the fall of Louisbourg, not from the east. The shifting shoals, the tideraces and currents of the mighty St. Lawrence were reckoned to be capable of defeating any foreign shipping without local pilots. With superb ship-handling, Vice-Admiral Charles Saunders' fleet of 170 vessels, of which twenty-one were line-of-battle ships, achieved strategic surprise for the youthful Major-General Wolfe's modest Army.

While the French hurriedly summoned reinforcements, until their forces about Quebec swelled to 5,000 regulars, 11,000 militia, 1,000 seamen and 1,000 Indians under the Commander-in-Chief himself, the very able Marquis de Montcalm, Wolfe set about disembarking and shaking out his expeditionary force of 8,500. By June 26, the whole fleet, manned by 13,750 seamen and marines, had arrived at the anchorage and base selected on the south-east corner of the island of Orleans, which lay mid-stream in the great river before it narrowed to go past Quebec, situated on its northern side on a plateau at the top of a forbidding escarpment; three companies of the 62nd, with Major Thomas Hardy as senior officer, were borne in Vice-Admiral Philip Durell's squadron; the fourth that had served at Louisbourg had sailed with their ship, H.M.S. *Burford*, to the West Indies. Below the main walled city lay the Lower Town by the St. Charles River, and across the St. Charles stretched some miles of the northern shore of the St. Lawrence, initially of mud flats and then small beaches and rising ground, to the Montmorency River. Here Montcalm disposed his main force of 12,000, with his own headquarters at Beauport.

With other marines, the 62nd disembarked thankfully from their fusty and confined shipboard quarters and enjoyed stretching their legs in the village of St. Laurent, among the gardens and small holdings and the setting of the great pine woods. They were soon loaded in the specially-constructed flat-bottomed boats and, rowed by seamen in these early landing craft, made a fresh and unopposed landing at the western point of the island which looked across two miles of open water to the imposing frontage of Quebec. Here a camp was set up, garrisoned by the 62nd under Major Hardy; in an old engraving from 'an authentic plan drawn by a Captain in His Majesty's Navy', this western tip is clearly titled 'Major Hardy's Post'.

From this point, also, Wolfe studied his objective. He drew little encouragement from the strong, natural position of the city and Montcalm's concentration of force at the only accessible landing areas nearby. However, he soon established Brigadier Monckton's brigade and several batteries at Pointe Levis, on the south bank of the narrowed St. Lawrence opposite Quebec and began a harassing bombardment. On July 8, he made a further and more drastic re-deployment. Leaving the 62nd and the other marine regiments to protect the Orleans base, he sent his other two brigades, Townshend's and Murray's, across to the northern bank of the St. Lawrence to dig in on the east of the Montmorency facing Montcalm's camps.

On July 18, an event significant to the fate of Quebec occurred. Covered by the artillery on Pointe Levis and under the cloak of darkness, H.M.S. *Sutherland* and some smaller vessels slipped past the guns of the city and proceeded upriver to destroy small French craft there and anchor blatantly.

At the end of July, however, Wolfe resolved to make a more positive move and break the deadlock: he would envelop Montcalm between a riverborne assault, using Monckton's brigade from Pointe Levis, and the two other brigades attacking across the Montmorency. As part of the build-up for this offensive, he moved two of the three 62nd companies from Orleans island to the Montmorency to protect the artillery redoubts.

The attack went in on July 31 and failed dismally, despite heavy fire support from the *Centurion*, a sixty-gun ship which normally bore a detachment of the 62nd, and two floating fourteen-gun batteries, and also the concentrated artillery on the Montmorency.

August was to be a month of severe strain for both British and French, however. With the important St. Lawrence river communications cut by British frigates above Quebec, food and gunpowder were scarce in the city and the camps, and though the French had incurred less casualties, their hospitals were now filling with sick. Moreover, the city was becoming scarred with ruins from the fire of the eighteen, twenty-four and thirty-two-pounder batteries on Pointe Levis. On the British side, some 854 casualties had been incurred, but, also, its field hospitals on Orleans were becoming crowded with sickness cases. More ominously, the season was drawing on and even the wholly co-operative Saunders had to warn Wolfe that the Fleet would soon have to depart.

Locally, the patient Major Hardy, the guardian of Orleans island, and the remaining company of the 62nd with him saw some welcome operational activity. Hardy and his company marched with a force of 1,200 under Brigadier Murray that moved along the south bank upriver, crossed in a specially-positioned flat-bottomed flotilla and conducted a large scale raid with mixed success. But at the end of August, Wolfe was ill with exhaustion and anxiety and morale was low in the British lines. If the French could hang on until the onset of winter, the whole British enterprise was lost.

However, on August 31, Wolfe was out of his sickbed, though still not fully fit, issuing orders for the re-grouping of the Army. With great skill, the Montmorency positions were evacuated and the brigades assembled at Pointe Levis for the march upriver along the south bank, with the heavy stores going by water. The three companies of the 62nd re-assembled at Hardy's Post, on the western tip of Orleans island looking across at the fine city of Quebec, ready to play a supporting role in the great operation about to unfold.

The brilliant feat of arms whereby Canada's fate was decided on the Plains of Abraham outside the city on the morning of September 13 was the fruit of a classic piece of military audacity: Wolfe, accompanied by his brigadiers, all disguised as grenadiers, spying the vital foot-track from a small cove that led to the escarpment only one and a half miles from Quebec while on his final reconnaissance of the north bank in a boat and his great decision to pin all on this hazardous one-man front to gain the plateau; the well-organized silent movement of 4,500 soldiers prepared for battle in ships' boats drifting downstream on the ebbing tide from Cap Rouge in the darkness of the night September 12/13 and surviving two French challenges ringing out from the north bank *en route*; the light infantry platoon scaling the cliff face through rock and scrub to rush the French picquets at the top of the track; the crucial disembarkation in darkness and movement up the track one man at a time; the deployment of British scarlet in two lines stretching one mile across the Plains as the sun warmed the morning air after the dawn; that terrible crashing volley delivered with two balls rammed into each Brown Bess musket that tore to pieces the attacking French ranks, and the charge with bayonet and claymore. In 1756-63, as in other wars, the hallmarks of the British soldier were his stolid fortitude, his excellent fire discipline and the accuracy of his fire. At Quebec, on that fateful morning, these sterling qualities were never more effectively to the fore.

The three companies of the 62nd took part in the deception aspect of the master plan that Vice-Admiral Saunders was to execute. On September 12, the previous morning, Saunders, with the bulk of the Fleet, weighed anchor from the Orleans base and moved across the great river to range threateningly opposite Montcalm's positions about Beauport; he began a desultory bombardment and lowered his boats crammed with the 62nd and other marine regiments. However, though this demonstration went on for most of the day, the astute Montcalm was not wholly convinced and he spent an uneasy night. At dawn next morning, he despatched an officer to the Governor of Quebec for any intelligence of happenings beyond the city over the past twenty-four hours, and then, at six a.m., unable to contain himself, rode off in person to Quebec. At the St. Charles River, he sighted his worst fears: the ominous line of British red on the heights. He summoned his regular regiments from Beauport, aligned himself at their head and went to his gallant death, as did his worthy adversary, Wolfe.

The 62nd disembarked to participate in the final act of Quebec's dying resistance. After the Fleet had closed on the Lower Town and the 62nd and other marines moved through the ruins to invest the city on yet another quarter, Vaudreuil, the inept Governor-General, surrendered when the militia refused to turn out as the drums beat to arms.

With the cessation of local hostilities, drafts from the 62nd and 69th, both marine regiments, were used to fill the depleted ranks of regiments who had suffered badly in the last and decisive battle.

A handful of officers only, due to these emergency drafting measures, returned to the United Kingdom with Saunders' fleet and these were too late to join the truncated Regimental Headquarters when it moved to Ireland in January, 1760. Here Lieutenant-Colonel John Jennings and his rather forlorn headquarters were joined by the surviving veterans from H.M.S. *Burford*, some three officers and forty-one men, who had been in action the previous year in the West Indies during the capture of the French islands of Martinique and Guadaloupe. With this valuable infusion of old soldiers to give steadiness to young Irish recruits, these elements of the 62nd, now nine officers and 130 men, were garrisoning the derelict fortifications of Carrickfergus Castle, on the north-east coast, when on the morning of February 21, three French men-o'-war sailed into the bay.

The French raiding squadron, under Commodore Thurot, promptly landed 800 regular troops to advance on the walled town and old castle. Throughout the day, fierce street-fighting raged about the town gates, through the cobbled alleyways of the town and to the breached walls of the castle as the greatly outnumbered 62nd made a desperate, delaying defence. Legend has it that as they expended all their ball ammunition in the last hours in the castle, they tore their round metal buttons off their tunics, rammed these into their smooth bore Brown Bess muskets and fired at the French. By late afternoon, out of ammunition and powder and about to be overrun finally, Jennings, also wounded, capitulated his detachment on terms. For minimal casualties themselves, the 62nd had inflicted losses of 50 dead and many wounded on the attacking French.

After re-provisioning, Thurot sailed away but was pursued by three British frigates and destroyed with con-

The Cup presented by the Mayor and Corporation of Carrickfergus to Lieut. Benjamin Hall of the 62nd Foot, 'in token of their gratitude and respect for his military skill, bravery and attention to their security'.

siderable loss of life off the Isle of Man two days later. The 62nd received a vote of thanks from the Irish Houses of Parliament for their gallant defence of Carrickfergus and it is thought that for many years they wore a 'splash' or indentation on their buttons to commemorate their last defiance to hold an overwhelming enemy.

The 62nd remained in Ireland throughout 1761 and 1762. As Colonel Kenrick writes:

> 'Ireland was one of the British Army's most fertile recruiting grounds, and the Regiment sorely needed men.'

Chapter 2

Return to the New World

To the eighteenth-century soldier, the threat of service in the West Indies hung like a death sentence. In those early days of elementary medical knowledge and crude sanitation, the beautiful islands of the Caribbean had a sinister and baleful reputation. Yellow fever, probably brought in by the slave ships from Africa, was the mass killer, aided by malaria, yaws and hookworm. Convicted criminals and captured deserters were offered reprieves if they volunteered for the insatiable West Indian drafts.

Among the West Indian spoils that fell to the British from the victories of the Seven Years War, the culmination of Pitt's brilliant maritime strategy, were the Leeward Islands. In the same year that the war ended, 1763, the 62nd Foot, now as a complete battalion, embarked in Ireland for service in one of the Leeward group, Dominica.

The Leewards, and especially Dominica, had an interesting and turbulent history. By the mid-eighteenth century, they had become the last strongholds of the Caribs, fierce, untamable cannibals who had moved north from the Brazilian mainland in a mass canoe migration via the stepping stones of the Leewards to raid and terrorize deeply into the Caribbean.

Another potentially dangerous force were the Maroons, so called from the Spanish *cimaronnes*, or wild men of the mountains, who were the descendants of escaped slaves from the early Spanish era inter-married with local Indian women of the peaceful Arawak tribe or kidnapped negress slaves. Having achieved a wild and lofty independence, the

Maroons cultivated a deep contempt for their very near kin, the negro slaves, and turned their detestation to some profit by running down fugitive slaves and returning them to their masters; in Jamaica, in 1760, the Maroons played a major part in the successful suppression of the greatest slave revolt in the colony's history. But they were a volatile and smouldering menace. In 1796, the 62nd were to meet the more numerous Maroons of Jamaica in the last, and most serious, Maroon War.

After five years of soldiering in a tense internal security situation, the 62nd, weak and depleted by sickness casualties, returned to the green fields of Ireland in 1769. This reasonable tour of duty in the West Indies was a remarkable feat, whether through accident or design, by the rudimentary military administration in the Horse Guards who, normally, were quite incapable of an efficient relief system. The 38th Foot languished in the Indies for sixty years.

When the American Revolution erupted in 1775, the 62nd was still serving in the Irish garrisons. As the news worsened, with the invasion of Canada by New England rebel militia in the autumn and the evacuation from Boston by Sir William Howe in the following spring, Lord North's uninspiring Government began despatching more and more reinforcements. In April, 1776, the 62nd embarked at Cork for active service in the Americas.

The actual destination of the Regiment was their old battle-ground of Quebec, where the Governor, General Sir Guy Carleton, who had been Wolfe's Quartermaster-General in 1759, had just repulsed a fresh New England invading force under the very capable Benedict Arnold. When the convoy bearing the 62nd and five other regiments arrived, Carleton followed up the retreating Arnold and inflicted a sharp defeat on him at Three Rivers, on the St. Lawrence route to Montreal; in this first action, the 62nd had a handful of wounded only. Carleton now drove the invading Americans from Canadian territory with two columns, a western thrust under Colonel St. Leger to Lake Ontario and a southern and main advance under Major-General John Burgoyne on the traditional route to Lake Champlain. The onset of winter and the cessation of active campaigning saw the British in command of the northern head of the strategic lake, poised for a counter-invasion of the rebel New England colonies.

By June, 1777, Burgoyne's army of 7,000, with Canadian boatmen, Indians, commissary and camp followers, had assembled at Fort St. John and Cumberland Point. Though Burgoyne, like many British commanders, had a poor opinion of colonial American military ability in comparison with veteran British regiments, he was not unaware that wholly European tactics would not succeed in the vast, undeveloped areas of America; as well as his recent Canadian experiences, he had fought in New England in 1775. He again grouped the light companies of his seven British regiments into a light infantry battalion; likewise, the grenadier companies into a battalion. Brigadier Hamilton's brigade, in which the 62nd served, with the 20th and 21st Foot, was also organized as light infantry, and the 62nd soon received the soubriquet of 'the Springers' for their quickness of movement.

On June 21, the advance began. Christopher Ward, in his *The War of the Revolution*, has given a vivid impression of that historic scene:

> 'Against a setting of the blue waters of the lake and the dark green background of its forested shores, the painted faces and bodies of hundreds of Indians, the masses of British scarlet and of German blue, the green of the jagers and the light blue of the dragoons, with their regimental facings of every hue, the shining brass of the tall hats of the Hessian grenadiers, the glinting of the sunlight upon polished musket barrels and bayonets, the flashing of thousands of wet paddles and oars made up a spectacular pageant, brilliant in its colour, light and motion, thrilling in its purpose and intention.'

By July 1, Burgoyne was somewhat cautiously investing Ticonderoga, whose fort commanded the southern and narrowed part of Lake Champlain, South Bay.

The British night occupation in strength of Sugar Hill, in the south, caused the Americans to abandon Ticonderoga. During the night July 5/6, boats containing important stores slipped off south to Skenesborough, while the American militia moved over a bridge of boats to Fort Independence and into the forest to retreat to the same destination via Castleton on the east. Brigadier Fraser, whose patrols detected the American withdrawal, took up the pursuit with the 24th Foot, the Grenadier Battalion and the Light Infantry Battalion, which included the 62nd's light company.

At 3 a.m. the following morning, he made contact with the American picquets. A fierce, running fight ensued, with the grenadiers commendably getting behind most of the rebels and seizing a steep hill over which the forest road ran in the area of Huberton, cutting off the retreat of some 1,500. The 24th and the light infantry then moved through the closely-timbered country to attack, and came under heavy fire. The action was finely balanced until a German force arrived, having followed in Fraser's wake in support. For 140 British killed and wounded, this successful follow-up action yielded some 200 American dead on the field, including Colonel Ebenezer Francis of the 11th Massachusetts Regiment, and 250 prisoners, with another commanding officer, Colonel Nathan Hale of the 2nd New Hampshires.

Among the officers present at the Huberton action was Ensign Thomas Anburey, 24th Foot, who describes in one incident the danger of fighting in close country:

> 'After the action was over, and all firing had ceased for near two hours, upon the summit of the mountain which had no ground anywhere to command it, a number of officers were collected to read the papers taken out of the pocket-book of Colonel Francis, when Captain Shrimpton,

of the 62nd Regiment, who had the papers in his hand, jumped up and fell, exclaiming he was severely wounded; we all heard the ball whiz by us, and turning to the place from whence the report had come, saw the smoke: as there was every reason to imagine the piece was fired from some tree, a party of men were instantly detached, but could find no person; the fellow, no doubt, as soon as he fired, had slipped down and made his escape.'

He goes on to record, as they erected log works and prepared to lay on their arms that night, the ever-present problem of supply in that undeveloped terrain:

'Provisions being unable to be forwarded, on account of the country being very hilly, a detachment was sent to shoot some bullocks that were running in the woods; these were distributed in rations to the men, which they ate, dressed upon wood ashes, without either salt or bread.'

On July 10, all Burgoyne's army, less a strong garrison of 900 he had left on the vital artery of Lake Champlain at Ticonderoga, had concentrated in the Skenesborough-Fort Anna area. Burgoyne had justifiable cause to congratulate himself; his advance was going well, his troops were flushed with success and now several hundred American loyalists joined him, despite the revolutionary committees threatening death.

Burgoyne now advanced through the forest to Fort Edward, on the upper reaches of the Hudson. It was a nightmare operation, not from direct contact with the rebels, but from the exhausting and continual work of clearing great trees felled by their expert axes across the straggling road, of building some forty bridges destroyed over creeks and of struggling through swamps; one marsh alone required a plank road two miles long. On July 30, he reached Fort Edward, having taken twenty gruelling days to cover twenty miles.

The British advance was now transfixed at Fort Edward by the mounting supply problem. The overland march left behind the boats needed for the Hudson, and the animal pack transport was quite inadequate to compensate; also, the artillery had to be brought forward laboriously.

Burgoyne was now acutely conscious of the rebels gathering about him in force, particularly on his slender lifeline to Canada. He set about gathering provisions for thirty days from the countryside, which took him until mid-September. Then, on September 13, his army, weakened by casualties and his garrisons to a strength of 5,000, crossed the river by a bridge of rafts to Saratoga. Heavy rain kept him encamped at Saratoga until September 19, when he moved out down the road by the river to the rebel main positions at Stillwater.

The Americans, numbering 5,000 Continental troops and 9,000 militia, were commanded by Horatio Gates, an Englishman who had served in the British Army under Cornwallis in Nova Scotia 1749–50 and the ill-fated Braddock at Fort Duquesne in 1755; he was a stolid, complacent commander but, unfortunately for the British, he now had Benedict Arnold as his second-in-command. His forces were entrenched on Bemis Heights.

Burgoyne, as he studied the ground, saw that a dominating hill on the left of the American main positions had only been partially dug and was certainly unoccupied and based his tactical plan on seizing that hill to roll up the enemy left. His attack was to be in three main thrusts: left, along the road by the river, von Riedesel and the Germans with Major-General Phillips, of 1758 Minden fame, and the artillery; centre, but moving on the unoccupied left hill, James Hamilton's brigade of the 20th, 21st and 62nd, with also the 9th, under Burgoyne; right, and moving deeply on to the enemy's left, Simon Fraser with the grenadier and light infantry battalions and the 24th. The terrain was typically difficult, being cut by several watercourses and was heavily-wooded; also, intervening high ground, on which Freeman's Farm was situated, lay in the path of the centre column.

The American outposts soon detected Burgoyne's own approach march. While Arnold characteristically wished to move out to attack the movement in its course, Gates, knowing the disciplined cohesion of the British regulars in the open field *vis-à-vis* his own volunteers, was inclined to wait behind his defences and let it develop. At 1 p.m. Burgoyne fired a signal gun three times to start all three columns from their assembly points. One hour later, after a steady advance through the thick woods, his own column closed with and drove in the rebel detachments holding Freeman's Farm.

Meanwhile, Arnold was able to make a sortie in force from the main positions and he attempted to work around Burgoyne's right to restore the Freeman's Farm situation. But *en route* his column ran into the fire of Fraser's units, undiscovered on rising ground on the right flank, and he had to break off with serious losses.

He now vented all energy, summoning reinforcements up to 3,000, on a frontal destruction of Burgoyne's four regiments now deployed in battle formation in Freeman's Farm. The woods about the Farm were soon swarming with rebels and for four hours, until the respite of sunset, Burgoyne's centre column, with the 20th on the left, 62nd in the middle and the 21st on the right, and the 9th in rear and reserve, withstood the onslaught of determined American attacks. To their aid came the indefatigable Phillips with four guns, and this invaluable artillery support, served by forty-eight artillerymen, of whom thirty-six were to fall, was a major factor in keeping the overwhelming enemy numbers in disarray. Eventually, as the long afternoon wore on the hard-pressed British regiments still stood their ground with disciplined musket fire and frequent bayonet charge, von Riedesel and some of his Germans came up on Arnold's right flank, having turned in from the river road axis, and the frustrated Arnold retired. As darkness fell, the weary and depleted British entrenched on the field of battle. Both antagonists had lost about 500 men, but the British, with their modest force, could not afford to suffer such comparatively severe casualties; though they had broken into and held part of the American defensive system, the main rebel positions were still intact.

At the close of the battle, Burgoyne rode along the decimated ranks of the 62nd, where only five officers and sixty men stood fit for duty, and said warmly: 'Well done, my brave Springers! . . .' But perhaps Thomas Anburey again more graphically sums up the heroism of that day:

> 'Lieutenant Hervey, of the 62nd, a youth of sixteen and the nephew of the Adjutant-General, of that name, received several wounds and was repeatedly ordered off the field by Colonel Anstruther, but his heroic ardour would not allow him to quit the battle while he could still stand and see his brave lads fighting beside him. A ball striking one of his legs, his removal became absolutely necessary, and while they were conveying him away, another wounded him mortally. In this situation, the Surgeon recommended him to take a powerful dose of opium, to avoid seven or eight hours' life of exquisite torture; this he immediately consented to do, and when the Colonel and Major Harnage, who were both wounded, asked whether he had any affairs they could settle for him, his reply was that being a minor, everything was already adjusted; but he had one request, which he had just enough life to utter: "Tell my uncle I died like a soldier".'

On October 7, Burgoyne resolved, with commendable courage and determination, to move out and attack Gates and Arnold again. With his dwindling force, he could have little hope of defeating the numerous and well-entrenched Americans and continuing on down the Hudson to meet up with Major-General Clinton, who was striking north from New York with 3,000 men: also, retreat to Canada was now a rapidly-shrinking possibility.

In this new attack, Burgoyne again strove to turn the enemy left. Directing the few Indians still remaining with him and some provincial sharpshooters to harass the rebel

3

flanks, he marched out of his entrenched camp with the 24th Foot, the light infantry and grenadier battalions under Fraser and the Germans under von Riedesel, a mere 1,500 men, and ten guns under Phillips; it is significant that the 62nd and the two other battered regiments of Hamilton's brigade were left to give local protection to the camp. Inevitably, the attack was too weak to make any impression and soon Gates began enveloping it with his own great numbers. The gallant Fraser was mortally wounded and the whole force began to struggle for its survival as it came under severe pressure. Burgoyne could only order a general withdrawal. With disciplined movement, the British fought their way back to the lines of the Freeman's Farm camp, sustaining many casualties and losing six guns.

Once again within the defences of their camp, the ordeal of the British was still not over. The ubiquitous Arnold had now placed himself at the head of a massive counter-attack which swarmed about the British positions. Serious penetration occurred in the frontages held by some Canadian irregulars and a weak Brunswick battalion but this was checked with much heroism by the light infantry under the Earl of Balcarres and the other German regiments; Breymann, the commander of the German Grenadier battalion, was killed. That night, Burgoyne withdrew his badly mauled regiments to further high ground overlooking the river.

Daylight revealed the Americans establishing themselves about him. He now made the bitter decision to abandon 500 wounded and sick and retreat upriver to Saratoga while he could still get through; there, with the Hudson protecting his left flank and a tributary his rear, he would await relief by Clinton. That night, in a storm with heavy rain lashing down, the tattered regiments trudged stoically and wearily to their final position. As they arrived there, men flung themselves down, exhausted.

The next day found the inevitable blocking movement by

the Americans, digging in on the opposite side of the river to prevent a crossing. Scouts also brought Burgoyne intelligence that rebel units were deployed in strength in his far rear between Fort George and Fort Edward, especially on the fords, finally crushing any slim hope of a forced march to the north without artillery or baggage, even if the vice of Gates' army, now swollen to 17,000, could be prised open. With some finality, Burgoyne was trapped. For a few days, he waited in vain for news of Clinton, but his men were now starving. On October 14, he began feelers to capitulate; on October 17, he surrendered with the honours of war.

Some 4,880 men surrendered at Saratoga, of whom, excluding American loyalists and Canadians, 1,900 were British and 1,600 Germans. The chivalrous treatment accorded them by Gates and his army was soon repudiated by Congress and Burgoyne's veterans disappeared into the oblivion of prisoners-of-war. Taken initially to Cambridge, Massachusetts, they were later marched south during the winter of 1777-78 to Virginia and suffered considerable privations in the cold and snow with little shelter and poor food. In 1781, the men were separated from their officers and vanished from the history of their motherland. The officers who survived captivity found their way back to England, some on exchange, including the commanding officer of the 62nd, Lieutenant-Colonel John Anstruther, and his second-in-command, Major Henry Harnage, both twice wounded in the fateful weeks between the battle of Freeman's Farm and the final humiliation at Saratoga.

In the campaign, the 62nd lost 316 battle casualties from an original strength of 541. In the high traditions of the British regular soldier, they gave their stalwart best, with the other Regiments, in a decisive and historic offensive which, if supported strongly by Howe from New York, could have changed the outcome of the whole war; its failure brought France, long resentful from the defeats of

Army Museums Ogilby Trust

'A Macaroni General', a satirical cartoon of Lt.-Gen. William Strode, one of the 62nd's early Colonels

the Seven Years War, into the struggle in early 1778, and in her train, Spain in mid-1779.

Battle honours, rightly, are not awarded for defeats and surrender. Yet the conduct of the troops under Burgoyne, soldiering in that magnificent wilderness against formidable odds, has drawn the especial comment of no less an authority than Sir John Fortescue. In his monumental *History of the British Army*, he states:

> 'In the whole history of the Army, I have encountered no grander display of steadfastness and fortitude than the heroic stand of the Twentieth, the Twenty-First and the Sixty-Second, with their little band of gunners, on the 19th of September.'

When the American war and the accompanying hostilities with France and Spain ended in September, 1783, a re-formed 62nd Foot was stationed at Trowbridge, Wiltshire. For the past three years, it had been built up in the Lincoln and Newark areas and during this period of rebirth, its flank companies did guard duty at Windsor Castle. Also in, 1782, under the new Army policy to link Line regiments with counties to encourage recruiting, the 62nd took Wiltshire as its territorial title.

After a year in Trowbridge, the 62nd (Wiltshire) Regiment moved to Ireland where they soldiered for six years. A tour of overseas service came in 1790, when the Regiment embarked for Nova Scotia. On arrival at Halifax, however, disembarkation was cancelled by a drastic change of orders. The 62nd was now to proceed to Jamaica.

The major core of violence and insurrection in the turbulent West Indies at this time was soon to be San Domingo, a large island near Jamaica and known in the old dominant Spanish era as Hispaniola. In September, 1791, a large-scale uprising of the negro slave population broke out in the French part of the island, now Haiti, and eventually, in some desperation, the French royalist government and planters appealed to the nearby British for protection. Accordingly, Major-General Sir Adam Williamson, commander of British forces in Jamaica, was given discretion to intervene and in September, 1793, he conducted landings at Jeremie and Mole St. Nicholas and marched inland to occupy other towns. In December, the flank companies of the 62nd were sent from Jamaica to Jeremie. For the next torrid six months, the flank companies, with other reinforcements mainly from England, fought in the operations that led to the occupation of the capital, Port-au-Prince.

But now the terrible yellow fever began to strike down the British invasion force and the death rate mounted appallingly. San Domingo was to become a grievous running sore. Over the five years of the British occupation, it has been estimated that the incredible number of 20,000 British troops died of yellow fever alone.

As the wasteful agony of San Domingo dragged on, the tinder fires of revolt began to spark in Jamaica. In July, 1796, two Maroons from their biggest settlement, Trelawney, were flogged for pig-stealing by a negro prison warder in front of assembled negro convicts and this humiliation was too much for the Maroons to bear, who delivered a virtual ultimatum to nearby Montego Bay Town that they would rise unless public redress and other conditions were met. The alarmed magistrates appealed to the newly-arrived Governor, Major-General The Earl of Balcarres; he promptly sent eighty dragoons, declared martial law and called out the militia. However, the crisis had caught him in a vulnerable situation: though a convoy of time-expired drafts from San Domingo was provisioning at the island before sailing on to England, his only full strength infantry battalion, the 83rd Foot, was under imminent orders for San Domingo and his two garrison regiments, the 60th and 62nd, heavily milked for drafts for the San Domingo war, could barely muster 300 men between them.

The 83rd duly sailed from Port Royal on the morning of July 29, but soon Balcarres yielded to his grave misgivings about their departure and despatched a fast boat to intercept their transports; this it did on August 2, at the northeast corner of the island and the recalled 83rd landed at Montego Bay. The following day, Colonel Sandford marched out from Spanish Town with 130 mounted dragoons and 100 men of the 62nd under Major Trevor Hull for the same rendezvous; Colonel Walpole, with 150 dismounted dragoons, embarked for the Black River to assume command in the St. Elizabeth and Westmoreland districts. Balcarres' outline plan was to ring in the rebellious Trelawney Maroons and intimidate them with a show of as much strength as he could muster. This had some mild effect when some thirty-eight Maroons, mostly elderly, surrendered in answer to his proclamation but the majority defiantly fired their township of Trelawney and moved into the inaccessible Cockpit country, wild, mountainous terrain in the nearby interior honeycombed with glens entered only by narrow defiles.

The retreat of the Maroon rebels to the Cockpit country was marked, however, by a sobering check meted out to the pressing Government forces. On August 12, a column under Colonel Sandford was ambushed and he was killed, together with thirty-six of his men. The Maroons also successfully attacked a picquet of the St. James's militia company of 'free men of colour'. The ambush was to be the pattern of Maroon tactics; agile and very fit, excellent marksmen and masters of bushcraft, they used the rugged Cockpit countryside with great defensive skill. Between ambush operations, they sent out raiding parties to terrorize nearby plantations and sow subversion among the slaves.

The end of August saw the start of the rainy season and the militia, unhappy in this uncomfortable and dangerous environment, melted away. On September 12, another severe local defeat was inflicted by the Maroons: Lieutenant-

Colonel Fitch, with men of his own 83rd Foot, probed too closely and ran into a well-sited ambush and he and some of his men were killed.

Though Balcarres spent as much time in the field as possible, the tense political situation, with the Jamaican planters and merchants fearful that the horror of another San Domingo was now in their island, forbade this as a permanency and the local Director of Operations, on the death of Fitch, devolved on Colonel Walpole, of the 13th Light Dragoons. To a dismal record of barely a known Maroon casualty against comparatively-heavy Government losses, he began to effect a gradual change by controlling the food-growing areas available to the rebels and driving them from water. The Maroons soon started to feel this pressure and became so desperate for water in their glens that they were dependent on rain catchments in the hollows of rocks and the bowl-like leaves of the native wild pine. They made one last serious foray out into the St. Elizabeth area, burning and murdering, but this time both sides had equal losses.

Meanwhile, Balcarres had sent to Cuba for a notorious local breed of mastiff used to run down fugitive slaves and bandits. Soon 100 of these bloodhound-type dogs and forty Spanish handlers arrived in Jamaica and the news of their impending deployment sent waves of terror through the normally arrogant and fearless Maroons. Almost immediately, though also the troops, with the company of the 62nd well forward, were now exerting strong pressure on them, the rebels began feelers for surrender. In January, some 400 rebels surrendered but as there were still a number of recalcitrants at large, Balcarres ordered the troops and the dogs to hunt them down.

By March, the revolt was over and the 62nd became fully occupied in the aftermath of the rebellion escorting and guarding Maroon prisoners.

In March, 1797, the remnants of the companies of the 62nd in San Domingo had sailed thankfully for England,

and the following year, the semblance of a half-battalion in Jamaica also embarked for home. In this typical West Indian tour of duty, the Regiment lost twenty-five officers and almost its entire original strength in men, mainly from the ravages of disease. In the drawing rooms and taverns of Exeter, the pleasant station where the Regiment now came together as a whole battalion at long last, the jaded veterans of the distant and dreaded Caribbean doubtless soon heard of a youthful French revolutionary general called Bonaparte who, that last winter, had won a campaign of brilliant and meteoric movement in the mountains and plains of North Italy.

Chapter 3

The Fight Against Napoleon

IN 1798, it was again time for another fitful expansion of the British Army, typically neglected since the end of the American War of Independence. The power of revolutionary France was rising to alarming magnitude and the serious rebellion in Ireland of the United Irishmen under Wolfe Tone, with belated French armed intervention, had been barely subdued. The next year, the 62nd Regiment marched from Exeter to Poole, Dorset, where it formed a second battalion, mostly of Wiltshiremen from the Militia. Militiamen, as a source of manpower for the Regular Army, were regarded as very sound material.

The tramp of regiments in camp and on the march and the bustle to build and refit ships clearly showed Britain's determination to face Napoleon Bonaparte and an erupting France, and the mood was infectious. Many young men flocked to the Colours in the patriotic fervour, but it was a hard, remorseless life. As one recruit of this period recounts, and as many of the raw Wiltshire lads in the 62nd's second battalion would have found:

> 'I remember well what sixpence a day was, recollecting the pangs of hunger felt by me . . . Of my sixpence, nothing like fivepence was left to purchase food for the day. Indeed, not fourpence. For there was washing, mending, soap, flour for hair powder, shoes, stockings, shirts, stocks and gaiters, pipe clay and several other things to come out of the miserable sixpence! Judge then of the quantity of food to sustain life in a lad of sixteen, and to enable him to exercise with a musket (weighing fourteen pounds) six to eight hours a day.'

Some reforms, however, were tardily in train, at least regarding dress. In 1796, the cumbersome skirted coat was abolished for a short jacket and the waistcoat also went. The cocked hat and tight breeches were to remain for over another decade, though the grenadier company in each infantry battalion now had distinguishing grenades on their epaulettes and the light company a hunting horn. The equipment of each man was a pack which, unsupported, cut uncomfortably into his armpits and crossbelts that bore a ball bag on the left, with a double frog for bayonet and hatchet, and powder horn on the right. The sergeants carried a fusil, an expensive, refined flintlock musket with a rifled barrel, and the officers were armed with a curved sabre.

In 1800, both battalions of the 62nd proceeded to Ireland, that sad, charming country with the tragic love-hate relationship with Britain: so bountiful with recruits and her blood for the British Army but so deeply conscious of grievances. Only two years later, the new-born 2nd Battalion was disbanded, a premature decision that was soon highlighted by another Irish rebellion, the Robert Emmet rising in 1803, which actively engaged the 1st Battalion. However, the sanguinity of such disbandments was even more exposed by the ominous threat of Napoleon Bonaparte assembling a vast camp at Boulogne, with accompanying invasion preparations. Again, the 2nd Battalion was raised, at Devizes in 1804, and as befitted the 62nd (Wiltshire) Regiment, was composed wholly of Wiltshiremen.

Meanwhile, the 1st Battalion, soldiering on in unhappy Ireland, was now beginning to be caught up in wider issues. In 1805, it was warned for operations in the West Indies, then to join the continental army at Hanover; in 1816 more abortive orders came for Portugal, but at last, in mid-September, the battalion sailed as part of 5,000 reinforcements for Sicily. The Commanding Officer was Lieutenant-Colonel Trevor Hull, who had commanded the 62nd detachments in the Maroon War in Jamaica.

A brief acquaintance with active service soon followed. A large British force had been deployed in Sicily since March 1806 to defend the island and the Neapolitan Court from the French in Italy, but one year later, a British expeditionary force also was futilely enmeshed in Egypt, in the hinterland about Alexandria.

The British return to Egypt in the spring and summer of 1807, after the glories of 1801, was not a rewarding experience. As Alexandria was alleged to be starving, British columns set out to secure the nearby food-growing areas of Rosetta and Rahmanieh. Two disasters occurred in succession in the Rosetta area; in the second battle, some thirty-six officers and 780 men, mainly of the 35th and 78th Regiments, were cut off at El Hamid by Mohammed Ali's troops and one-third killed, the others captured with many wounded. Though the Mamelukes moved north to support the British, by May all the British Commander could think about was the safe recovery of his missing and extrication from the whole, useless mess.

On May 25, valuable sustenance came to the British enclave with the arrival from Messina of 1/62nd and 1/21st, brigaded under Major-General John Sherbrooke, and two supply ships. The 62nd moved upcountry in the blazing heat of the Egyptian summer to the Rosetta-El Hamid area. For three months they acted as fresh infantry to counter the pressures of a confident and numerous enemy. In September, the 62nd took a major part in covering the withdrawal, to which Mohammed Ali yielded his prisoners who had been, surprisingly, well-treated, and on September 19, the British rearguard sailed off to Sicily.

When the sorry Egyptian expedition returned, 8,000 of the best British troops in Sicily left for Gibraltar under Sir John Moore and the difficult Sicilian command devolved on the 62nd's brigade commander, Sherbrooke, a rough, forceful Indian veteran. With some 7,000 British and foreign soldiers, he had to contend politically with the intrigue-ridden Court of Naples, subsidized in refuge at Palermo with British money, and which was detested throughout Sicily and Italy for its corruption and laxity; the surviving Neapolitan forces and Sicilian militia, both badly-equipped and ill-administered, did not come under his command. On the mainland, the enemy was personified by Joseph Bonaparte, King of Naples, and throughout Italy Napoleon's eagles were supreme, except in the wild fastnesses of Lower Calabria where gangs of savage *patriotti* cut down isolated French detachments and couriers. Here, across the Straits of Messina, the British and Neapolitans still held several strategic fortresses.

It was at such a fortress, the castle of Scilla on the northern tip of the Straits, that the 62nd was soon in action again. The Peace of Tilsit between the French and Russians set reinforcements flooding in to Joseph Bonaparte and he resolved to clear Lower Calabria to establish a mounting area for the long cherished invasion of Sicily and expulsion of the British. In December, 1807, the French Major-General Reynier pressed south with a division and set about investing Scilla with five battalions and some cavalry, though he had no heavy guns. Scilla was held by 400 British troops under Major Robertson, 35th Foot, rated an excellent soldier: a company each from 1/62nd, the 27th and 58th Foot and artillerymen. The castle was sited on a rocky promontory dominating the Straits and British sappers had cut steps in the stone of the seaward cliff-face as an emergency lifeline for supply and reinforcement and, ultimately, withdrawal to Sicily.

Throughout January the small garrison defied Reynier's assaults but on January 30, the fortunes of war changed drastically. Four Neapolitan gunboats sent over by the Military Governor of Messina to give bombardment support deserted to the French and Sherbrooke, who detected the treachery from his Sicilian observation point, furiously sent the British brig-of-war, *Delight*, in pursuit. Tragically, this ship in its zeal ran in too close and, her decks raked with French fire, went aground and had to be burned. From her, despite the gallant efforts of the sloop *Bittern*, who had sailed to the sound of battle, the French secured four twenty-four pounders. Three days later, Reynier gained more guns from the fortress of Reggio, when its Neapolitan governor surrendered.

> 'By the 4th, the serviceable guns from that castle, as well as the 24-pounders from the boats, had been transported to Scilla, and brought under cover of the houses to breaching distance; other batteries were armed with howitzers and field pieces, and the fire of the whole was directed at the fort. Major Robertson and his little garrison defended their post with admirable intelligence and bravery; but the match was too unequal. The French made one more effort to get round the foot of the rock and cut off the means of retreat afforded to the English by the steps from the sea: but these attempts cost them dear. At length after the castle had been pounded during thirteen days, its guns dismounted, and the wall facing the town beaten into ruins. Robertson found it necessary to evacuate the post he had defended so gallantly. Boats were collected silently in the night; all the wounded who could bear removal were carried off; and company after company descending the steps at the back of the rocks were embarked with little loss, the rearmost sections keeping up a fire upon the enemy. The brave garrison of Scilla rejoined their comrades at Messina on February 18, leaving a mere ruin to be occupied by the enemy.'

In April, 1808, Sherbrooke was relieved by Sir John Stuart, a strange, ineffective commander who had won the minor victory of Maida on the mainland in 1806. Under his colourless leadership, morale began to deteriorate, and experienced regimental commanders drifted away to England and Spain to find other employment. The cheap and strong local wine led to much brawling with the Sicilians, and the health of the army was affected by the malarial climate. One unhygienic facet of Army life, however, was now removed as the ponderous march of administrative progress moved slowly on: the powdered and pigtailed long hair was abolished at long last.

1808 had ended ignobly for the British in the Mediterranean with the dashing Murat, who had succeeded Joseph Bonaparte at Naples, capturing Capri; belatedly, in April the following year, Stuart bestirred himself to give some offensive action to his restive army. After much brooding over Calabria, he decided to take the two islands of Ischia and Procida in the Bay of Naples to threaten the city. On June 11, 1809, an armada bearing 13,000 troops, including 1/62nd, sailed for the Bay of Naples; a diversionary force parted company to retake Scilla Castle. The British appearance in force in such proximity caused great excitement in Naples and Murat hurriedly forcemarched his outlying divisions to the defence of his capital. But, apart from this rather negative success, the large British force sat on Ischia and Procida, with no local supplies and little water, under the indecisive Stuart, wondering, with him, what the next move should be.

But the great opportunity offered by the Italian peninsula to a fleet and army working closely together was passed over by Stuart who felt, on the contrary, that events directed retreat. The operation against Scilla had failed before the sudden appearance of a French division, which drove the British force to peremptory evacuation. The ensuing panic in Messina persuaded Stuart to embark his whole army to return to Sicily.

Peace with Austria set more French divisions on the march into Italy. After a French invasion fiasco south of Messina in September, 1810, which was soundly defeated the 1/62nd spent the remaining months of 1810 and all 1811 on coastal defence between Messina and Faro but this defensive existence was enlivened by occasional participation in amphibious raids.

In March, 1811, Lord William Bentinck succeeded Sir John Stuart as Commander-in-Chief but Sicily was now a backwater to the major theatre of operations, Spain. Like Stuart, he strove to resist War Office plans to transfer some of his forces to eastern Spain, but in November, 1812, he had to yield and despatch 9,000 troops, which included the grenadier company of 1/62nd, to Alicante to campaign in Catalonia against Marshal Souchet. When 2,000 men of this force returned to Sicily in March the following year, the 62nd's grenadier company was among them. Late 1813 saw both battalions of the 62nd in the Mediterranean, as 2/62nd came out to join the Peninsular Army. The two Commanding Officers had splendid Dickensian names, George Gauntlett with the 1st Battalion, Nathaniel Blackwell with the 2nd.

1813 was also a year of gathering strength for the British in Sicily. Lord Bentinck was ruling the island in all but name, and relations with the local people improved considerably. In early 1814, secret negotiations were opened with Murat, who was trying to salve his throne out of the coming wreck of the Napoleonic Empire, and Bentinck, in concert with the Austrian commander in Northern Italy, planned to land in force to capture Genoa. In March, 1814, 1/62nd sailed from Sicily as part of the assault 1st Division to Leghorn and when the transports returned with the 2nd Division, Bentinck advanced on Genoa. Two brisk actions, ejecting the French from Siestri on April 8 and from Nervi on the 12th, soon brought the British to the outskirts of Genoa and the 62nd became closely involved in confused fighting in the built-up suburbs. But by April 21, for a cost of 200 killed and wounded, the city was occupied.

However, as the soldiers of 1/62nd paused at Savona, dramatic orders came to embark for America. With the abdication of Napoleon, veteran regiments from the Peninsular Army and from the army in Italy were being transferred with some speed. This massive shift of British sea and military power soon began to tell and already, in the New England states, a vociferous peace party agitated for the end of the war that had now dragged on for three years.

In May, 1/62nd sailed from Genoa for Nova Scotia, where it came under the command again of its old general of Egypt and Sicily, Sir John Sherbrooke, the Governor.

In late August, Sherbrooke sailed from Halifax to invade Maine with ten transports loaded with, among others, the 1/62nd, reaching the mouth of the Penobscot River on August 31 where his naval escort was joined by the seventy-four-gun *Bulwark* and four other warships. Next day, as the 62nd and the other regiments prepared to assault, the commander of the American fort at Castine blew up his post and withdrew the garrison upriver.

At Castine, Sherbrooke learned that the American corvette, *Adams*, was lying up the Penobscot and sent off to locate and destroy her a task force of 600 troops, including men of the 62nd embarked in small vessels and barges. Upriver at Hampden, the British force was confronted by 600 Maine militiamen, about 200 seamen from the *Adams* manning a battery of ship's guns mounted by their captain on a high bank of the Soadanscook Creek which commanded the river approaches, and forty soldiers from the Castine fort. Next morning, the British regulars attacked and before the steady advance of the veterans, the enemy force disintegrated, the militia fleeing with hardly a shot fired. The British took eighty prisoners and twenty-five guns and pressed on to Bangor, where they received the surrender of an American general and 190 men; the *Adams* and two other ships had been burned.

On September 9, another column which included the other companies of the 62nd, disembarked at Buck's Harbour from warships to capture Machias. After a difficult night march, they came up in the rear of Fort O'Brien at daybreak but, again, the garrison fled without a fight. That

4

day, Machias and its Point battery were occupied and the whole of eastern Maine from the Penobscot to the New Brunswick border came under British rule.

However, apart from the battle of New Orleans, the war was almost over and in the peace terms, Maine was handed back to the Americans. The 1st Battalion left the soil of the United States and went into a garrison life in Nova Scotia.

Meanwhile, the 2nd Battalion had remained at perennial invasion stations in England until 1809 when it moved to the exposed garrison of Jersey. Ultimately, it moved back to Ireland, and it was there, 1,000 strong, in 1813 when it was placed under orders to join Wellington's army in the Peninsula.

After nine years of make-believe wartime soldiering, 2/62nd Regiment was now to join one of the more professional and closely-knit armies Britain has ever placed in the field. Of polyglot uniform due to the wear-and-tear of years of active service abroad, inured to battle, fit and capable of great endurance, driven to splendid achievements by either fear of the lash or devotion to respected officers and equally able to resort to drunkenness and looting when success broke down their iron battle discipline (until finally restrained by their officers and sergeants), Wellington's veterans were the epitome of stolid British courage and hardened field soldiers.

From the impetus of their years of campaigning, reforms, both tactical and administrative, came along relatively fast. The East India pattern musket, another modification of the tried Brown Bess, now gave a lethal range of 80 yards for the good shot, but its real tactical value was that a trained soldier could reload and fire again between twelve and fifteen seconds, while the Baker rifle, with which 4,000 men of the Allied Army was armed, though accurate up to 300 yards, took thirty seconds, a serious drawback which the great Napoleon, for one, refused to countenance; he withdrew all rifles from his armies. Artillery was making rapid progress, with canister shot reaching 200 yards and more against massed infantry or cavalry, and the round shot that could achieve a range of 800 yards and still went bouncing and careering murderously about; and shrapnel was an effective British invention. Overall, the use of black powder shrouded the field in a literal fog of war, even on sunlit days, a great problem as the Commander's battlefield intelligence depended on the efficacy of telescope and galloper. In the not-unimportant direction of attaining an efficient and practical fighting dress, the cocked hat yielded in 1811 to the Wellington shako and grey trousers and half-boots replaced tight white breeches and gaiters.

The 2nd Battalion disembarked at Dassages on October 6, 1813, to join Lord Aylmer's independent brigade. Wellington had now reached the Pyrenees with a British, Spanish and Portuguese army and was about to debouch on to the soil of France. Confronting him, on three major defence lines, lay Marshal Soult with a French army of equal strength.

At 7.15 a.m., October 7, Wellington attacked the first defensive system, the line of the Bidassoa River from the sea, with the main weight of his offensive on the left or sea flank. Here Sir Thomas Graham's corps, which included Aylmer's brigade, struck a weak sector held by only two French divisions as Soult expected the main attack further inland; his Intelligence reckoned the river below Ile de Conference towards the sea where it stretched out in a tidal estuary 1,000 yards wide and twenty feet deep was impassable. However, the local Basque shrimpers knew that at low tide it was fordable in certain places and after mid-September, so did Wellington. Middle-deep, the British infantry waded slowly across and soon a pontoon bridge was established. By noon, the attack on the left had carried itself forward over all its objectives and with Sir Thomas Graham planting the British flag on French soil, his divisions halted before the fortified lines of St. Jean-de-Luz with only 400 casualties.

Inland, stiff fighting occurred, with 1,000 casualties in the Light Division and a Spanish division, but as the day's action was broken off, the British were firmly consolidated on the right and centre of Soult's first line. In their baptism of fire, 2/62nd had participated in an auspicious success and on the crucial sector.

After the crossing of the Bidassoa, Wellington paused for a month to build up his supply position. The hardships of survival in those bleak mountains, amid the early snow and raw cold of mid-autumn, were severe. The soldiers suffered deeply and the toll of sickness began to mount.

On November 10, at 6.15 a.m., Wellington attacked to break into the French second line, from Jean-de-Luz on the sea inland along the River Nivelle. The left corps, with 2/62nd, was now commanded by Sir John Hope and it was to go forward in a general advance by the whole Army, but for this operation, the main effort was to be made in the centre by Beresford's corps. It was a beautiful moonlit, starkly clear and cold night as the Allied divisions tramped silently forward to their startlines. Then, after H-hour:

> 'It is impossible to conceive a finer sight than the general advance of our army from the Pyrenees passes against the French position. Almost as far as the eye could reach, was seen one sheet of flame and smoke, accompanied with an incessant fire of light troops and frequent volleys of musketry, as the lines and columns approached the entrenchments. . . . The troops on the left, under Sir John Hope, drove the enemy before them; and by their intrepidity and determination, overcame numerous obstacles, both natural and artificial, which they encountered in their advance. Marshal Soult withdrew his army across the Nivelle during the night, and destroyed the bridges over it near St. Jean-de-Luz, evacuating that seaport. The allied army bivouacked for the night on the heights which it had taken.'

The weather broke torrentially the following day, flooding the execrable country roads into deep mire and rendering any pursuit a laborious and slow operation. Gradually, a line of Allied positions was set up before Soult's final defence system, the principal strength of which was that the River Nive, at one stage of its course, would divide Wellington's army, exposing it to defeat in detail if Soult could concentrate rapidly.

On December 9, five Allied divisions duly crossed the Nive against light opposition only, but next day Soult counter-attacked with the main weight of his army between the Nive and the sea. Two French divisions drove in the weak outpost line of Hope's corps and quick to reinforce success, sent another division to exploit their penetration. This very ugly situation was saved by Lord Aylmer's brigade, forcemarching ten miles from St. Jean-de-Luz to take the French in the flank and throw them back.

After his near success on the left bank of the Nive, Soult countermarched and attacked Hill's corps on the inland sector. But his heavy casualty losses, together with the wholesale desertion of his German regiments, forced him to break off the battle and his demoralized army streamed away from the Nive. Soon he was assailed with demands for reinforcements from Napoleon defeated at Leipzig, and his command shrunk to 42,000 whereas Wellington mustered 60,000. 1814 showed the way to be clear for a general advance by the Allied forces deeper into southern France. The 2/62nd were at Bayonne when the war ended, with the eastern allies entering Paris on March 31 and the Emperor abdicating on April 6.

At the end of August, the Regiment embarked for Ireland where it landed on September 7 at Cork. The resurgence of Napoleon saw it sailing for Europe again, landing at Ostend on July 8, 1815, far too late for the glory of Waterloo, to march to Paris where it was quartered in the Army of Occupation until January, 1816. On the voyage back to Ireland, two companies were shipwrecked off Kinsale, but with little loss of life. Finally, on April 24, 1816, 2/62nd

Regiment was disbanded at Dublin. However, a large transfusion of its spirit and standards, in the persons of a draft of 400 men, went to the 1st Battalion, now back in Halifax.

So began the break-up or dispersion of the many veteran battalions who had fought so courageously and long during the Napoleonic wars:

> 'scattered all over the world, like dust before the wind. A few regiments went to England, many to Ireland and still more to the East and West Indies, Canada, the Cape, the Mediterranean; and, in short, to every creek and corner where we had colonies or garrisons. Thus was dispersed an army, many regiments of which had been together from the year 1808 to 1818.'

Chapter 4

'The Sun Never Sets'

FROM this scattering of the Regiments, the 62nd returned to England from Nova Scotia in 1823. After landing at Portsmouth for a fleeting month, it sailed on to Ireland and disembarked in late October to go into billets in the county of Tipperary. Over the next seven years, the Regiment served in the Irish garrisons, at Dublin, Enniskillen and Londonderry, recruiting hard to fill an establishment of ten companies, each of seventy-four rank and file. Then on March 22, 1830, the Regiment marched from Limerick to Cork to embark for Chatham. It had been ordered to India, to be stationed in the Madras Presidency.

In India, the octopus of British power was spreading irresistibly towards complete domination of the seething and colourful sub-continent. The Mahrattas had been defeated in 1817, and the Pindari scourge broken soon afterwards. Only the Sikh Empire of Ranjit Singh, from the Punjab to Peshawar, stood intact from the seemingly invincible combination of British military success and mercantile penetration.

The 62nd disembarked at Madras in October, 1830, and marched to Bangalore, in the Mysore Division. 1831 saw disturbances in the Nugger district of Mysore, and in one of those sudden minor operations that frequently arose in turbulent India, Major Singleton of the 62nd was ordered to take the grenadier and flank companies off to join a field force assembling under an East India Service officer to pacify Nugger. The flank companies returned to Bangalore two months later, having lost nine men but from cholera, not wounds in action. This was an ominous hint of tragic circumstances that were to dog the Regiment during its years of service in India.

In 1833, the 62nd was ordered to relieve the 45th Regiment at Masulipatam, one of the more unhealthy stations in southern India. Moreover, despite the protests of the Commanding Officer, Lieutenant-Colonel Thomas Reed, who sought to go by sea at least part of the way, the Regiment was to cover the 400 miles by march route. This journey developed into a horrifying death-march, as an old regimental account relates:

> 'The Regiment proceeded on its way on the 18th February and arrived at Chittoor, in the Carnatic, on the 1st of March, where it halted for a muster. On resuming the march, Indian cholera, which had been raging in the district, leaving whole villages depopulated, made its appearance among the troops, and the first victim was a poor woman, wife of one of the private soldiers. At Kulgerry, nine miles from Chittoor, several admissions into hospital took place; five of the men were buried on the evening of arrival, along with the child of the surgeon of the regiment, Dr. Radford. All the way thence to Nellore, where the regiment arrived on the 16th, the fatal disease continued its ghastly work, and its ravages increased among the inhabitants of the country as the troops. Each day added difficulty of procuring transport for the hospital, and this was only effected by daily pressing from the villagers on the route great numbers of half-starved natives, who were employed to carry the sick in blankets, suspended from the branches of trees, bamboos. All this while the cattle employed in the transport of the general baggage of the regiment were daily dying from fatigue and want of forage, the country being completely burnt up from the long and unusual drought.
>
> 'At Nellore, the regiment encamped on the left bank of the Pennair river, under the pleasant shade of a large plantation of tamarind trees; additional medical aid was received from Madras, and for the first time the cholera began to abate, and when the camp was broken up on the morning of 21st March, and the march resumed, it was

believed to have altogether disappeared; but it soon unfortunately returned in all its virulence. On the 26th, the regiment reached the sea, and encamped at Ramapatam, almost upon the beach, and upon the 27th marched with the sea on the right flank to Carvador, and on the 9th April arrived at Vellasoor, on the right bank of the Kistna, where the regiment halted upon the 10th, to complete the arrangements for crossing this rapid and deep river. As the termination of this unfortunate and melancholy march approached, admissions to the hospital from cholera became less, and on the 13th the regiment arrived at Masulipatam with 386 men fit for duty. It encamped outside the walls of the fort, upon the saltwater swamp which surrounds its walls, until the 16th, when the headquarters of the 45th marched out *en route* to Secunderabad; and the regiment marched into the barracks in the fort, still continuing sickly.

'On the 21st May the usual hot winds set in with considerable violence, and the admissions into hospital during the three following days were upon an average of 50 men per day, many cases proving fatal, principally from apoplexy and *coup-de-soleil*. At this time there were only 44 men out of hospital fit for duty; the garrison guards were discontinued and the regimental one decreased.'

In early 1834, the health of the Regiment improved temporarily until the rains in June when, again, fever and dysentery swept through Masulipatam and the 62nd began losing seven men a day. The Governor-General, Lord William Bentinck, the 62nd's old commander in the wartime Mediterranean, now took some firm action: he abolished Masulipatam as a station for British troops and ordered the transfer of the Regiment to Moulmein, Burma, where they were joined a month later by a bountiful draft of reinforcements from England. Two more drafts quickly followed, building up the 62nd to a reborn 698, and this splendid infusion of manpower, coupled with the pleasant surrounds of Moulmein, ensured recuperation from the horrors of Masulipatam.

After six years in Burma, the 62nd was ordered back to

Bengal; the situation on the northern frontiers, where the First Afghan War was in progress and the Sikh and Nepalese borders restive, was disturbing. A year was spent at Hazarabagh, a healthy hill station in Bihar, then the Regiment moved down to Calcutta to relieve the 50th Regiment at Fort William, which always had a Queen's regiment as garrison. The ceremonial and social activity of this seat of John Company's power, with its durbars and levees, and the earthy temptations of Calcutta had a debilitating effect on all ranks and it was doubtless with some relief to the efficiency of the Regiment that it was ordered to Dinapore in July, 1842.

The move was in boats on the River Ganges, and after embarkation of the first half of the Regiment on August 11, all went well until the afternoon of September 5, an ominously sultry day. At 2 a.m. the following morning, a hurricane blew up and capsized many of the boats which were driven violently from their overnight moorings. Many were drowned and the Colours, the regimental records and all the Officers' Mess silver were lost except one piece of silver: 'a silver snuff box which was used at the time by the Adjutant. This was the one presented by Lieutenant-Colonel George Gauntlett in 1811.'

The 62nd remained at Dinapore until relieved by the 39th Regiment in December, 1844. It now continued its Indian travels by marching off to Ambala, the main British military station confronting the Sikhs. For it was now apparent to the masters of British India that a trial of strength was building up with the formidable Sikh Army, and a discreet but positive concentration of Queen's and East India Company troops was ordered to the vicinity of the frontier, the River Sutlej.

Since the death of the great Ranjit Singh in 1839, the Sikh Empire was in some disarray. His successor, Shere Singh, had been murdered in 1843 and now the Rani, the mother of the present boy king, Duleep Singh, was encouraging the powerful Khalsa, the standing army, to enter into the traditional distraction of a foreign war to take their pressure from her son. The strength of the Khalsa varied between 60,000 and 100,000 well-equipped soldiers, with an excellent artillery arm of 200 guns trained by French and Italian mercenaries, whereas with the British the bulk of their forces were East India Company native infantry regiments and this was a worrying weakness.

On arrival at Ambala, the 62nd's destination was changed to Ferozepore, a forward outpost on the Sutlej. As 1845 dragged out its hot and dusty course in the Punjab, the Regiment camped in considerable discomfort, with the temperature soaring up to 120 degrees in that flat, semi-desert countryside. Meanwhile, both Sikhs and British prepared for the coming clash. The Commander-in-Chief, Gough positioned himself at Ambala, and was joined later in the year by the Governor-General Sir Henry Hardinge; at Ferozepore, Major-General Sir John Littler assumed command and organized his force into a division of two infantry brigades and one cavalry brigade. The 62nd, the only British regiment, was brigaded with the 12th and 14th Native Infantry under Colonel Thomas Reed, while another 62nd officer, Lieutenant-Colonel The Honourable Thomas Ashburnham, commanded the second temporary infantry brigade. As the year drew to a close and tension mounted with minor Sikh incursions, the 62nd was much occupied with active patrolling and standing-by to turn out at short notice.

On December 12, intelligence reached Ferozepore of crossings in force by the Sikh Army over fords and ferries on the Sutlej between eight and twenty miles away. Once over the river, the boldness of the Sikhs rapidly deserted them and they set about constructing a large fortified camp about the village of Ferozeshah. Littler marched his division out to block any further forward movement towards Ferozepore and to observe the Sikh build-up.

Meanwhile Gough and Hardinge were soon marching out from Ambala with the main British-Indian field force.

On the 18th, Gough was at Moodkee, his troops having trudged 150 miles through the dust and heat for six days, and found his route barred by an equal enemy force of 12,000 under Lal Singh. Darkness closed a bitter, rather indecisive battle, though Lal Singh did withdraw with the loss of a number of his guns. Gough camped on the field to regroup and bring up his ammunition train, and sent final operation orders to Littler who received these instructions at midnight on the 20th. He was to disengage from his watching position before Ferozepore and effect a junction with the main army at Sultan Khanwallah, ten miles to his east.

At 8 a.m. next morning, Littler slipped his division away unhindered by enemy columns by the simple pretence of leaving his camp standing, and from Moodkee, Gough also set out, leaving two sepoy regiments to guard his supplies and the wounded. By mid-morning he was near to the great Sikh camp at Ferozeshah.

As his weary men rested, Gough eagerly proposed that he should attack at once against the more open eastern flank of the Sikh positions, against which his army had opportunely arrived. But Hardinge insisted that Gough should first join up with Littler's force. It can only be a matter of conjecture whether Gough, against the general run of his ability, had hit upon a perceptive *coup d'oeil* in considering that the easier eastern perimeter offset his going on to meet Littler, or whether Hardinge's caution was rightly sound. Certainly the juxtaposition of the army after joining up with Littler meant that the plan of attack went against the strong southern defences, where the main weight of the enemy artillery lay.

Gough moved his army on to the south-west and at noon Littler's division came up. The whole army turned in to face the south of the Sikh positions and Littler's division moved into its battle position on the extreme left, deploying at the village of Misriwala. At 3.30 p.m. after hardly any co-ordination, the British-Indian army of 18,000 with sixty-five guns, all field pieces except two heavy howitzers, advanced to attack 30,000 well-entrenched enemy of a fearsome military reputation, supported by over 100 guns.

For some reason Littler's division was well ahead of the other two assault divisions. Thomas Reed's brigade, with the 62nd on the left was in the van, three sepoy regiments in support. Their axis of advance was through a brushwood jungle which made cohesion difficult and did not wholly conceal their approach from the observation of the waiting Sikh gunners. As their redcoated lines trudged on through the scrub to within range, the well-trained enemy gunners soon found the correct range after a few exploratory rounds, and the consequent persistent and accurate defensive fire from the Sikh guns soon began to wreak a mounting number of losses; the 12th N.I. ranks on the 62nd's right began melting away and the 14th in its rear dropped ominously behind.

Three hundred yards from the Sikh entrenchments, the 62nd emerged from the partial concealment of the brushwood into open, exposed ground. Into its long lines tore a most devastating fire of grapeshot and canister. The 62nd was literally blasted to a standstill as fearful gaps were blown in its sweating ranks. The carnage inflicted in its frontages in those horrific minutes as salvo after salvo roared from the Sikh positions, backed by musketry fire, was terrible. In this bloody field, the 62nd was alone and decimated, and squadrons of enemy cavalry began appearing on its left flank, like vultures for the kill. The sepoys in its own brigade and Ashburnham's had fled or not come forward, and away to its right, the other divisions were some distance behind.

With many of their officers and N.C.O.s crumpled in inert or writhing groaning heaps among them, the surviving, shattered ranks of the 62nd halted and began firing at the

Sikh positions. Thomas Reed, the acting brigade commander, eventually got the firing to cease, as the acting-commanding officer, Major Shortt, was among the wounded, and ordered the survivors to charge. But the men were incapable of covering those remaining yards of that fire-raked ground; shocked and dazed from the hail of fire that had seared through them, they were also physically exhausted from the approach march that long, dusty morning and the ensuing peremptory rush into battle. Some raggedly gained the Sikh trenches; but those who did were cut down with Sikh swords, Reed now made the only practical, hard, tactical decision: he ordered retirement. Even if the survivors had gained the Sikh positions they would have been quickly despatched, few, disorganized and isolated, by the packed Sikh infantry protecting the guns.

The remnants fell back through the protecting ranks of the 9th Foot. Sir Harry Smith had hurried forward as he predicted the debacle of Littler's abortive and isolated attack, and the whole of his left brigade was soon to become embroiled repelling a strong Sikh counter-attack on the left flank of the army. Now a tremendous battle belatedly developed over, and overlapping, the whole of the Sikh southern perimeter as the two other divisions, in a holocaust of dust and smoke and roaring artillery, closed with the enemy and went for their gunners and protecting infantry with the bayonet. Sir Harry Smith's other brigade passed through the two assault divisions embedded in the forward positions and began fighting through the village of Ferozeshah, while on the eastern perimeter the 3rd Light Dragoons made a brilliant charge, overrunning guns and sabring the gunners.

Darkness fell on scenes of indescribable confusion: of exploding magazines, blazing tents and buildings and mixed-up leaderless units of intensely-excited men. After the capture of Ferozeshah village, Smith found himself and many men lost almost in the middle of the Sikh position and in danger of being enveloped by the numerous enemy, despite his forming a defensive semi-circle; many of his sepoys began breaking when a heavy enemy gun started firing canister shot in their rear. At 2.30 a.m. he got his isolated and polyglot command under sufficient control to file away from the equally confused and excited enemy along the west side of Ferozeshah and eventually over the harrowing field where the many dead of the 62nd lay huddled in the cold night air.

A night of extreme tension and mutual apprehension settled over the two exhausted and disrupted armies. The situation was more critical for the British who withdrew for their own local security from the captured Sikh forward positions and razed Ferozeshah village and lay down on their arms near their old startlines: the handful of vital Queen's regiments, especially the 62nd, had all lost heavily. It was a night particularly burdensome with decision for Hardinge and Gough, and, it is perhaps easy to smile over the years at these two pro-consuls of empire, both former officers of the Duke of Wellington in the Peninsula, when considering the splendid tenor of the conversation reported to have been held that fateful night, but they were men of profound courage, moral as well as physical.

> ' "It is my duty to tell you," said Gough to Hardinge, "that the army is in a critical position. What do you recommend to be done?"
>
> "Wait until daylight, attack the enemy vigorously, beat him, or die on the field."
>
> "Agreed," said Gough, and they parted.'

At first light the next morning, Smith rejoined the main body of the army with his separated troops and the British prepared to attack again; in the hours of darkness, many of the Sikhs had filtered back to the vacated positions. On this second day, the remnants of the 62nd were mainly commanded by their sergeants, as eighteen out of twenty-three officers were casualties.

After three troops of Royal Horse Artillery had bounded forward to give vital fire support, the infantry advanced again into the attack with the bayonet. But the Sikhs had lost heart, probably much due to a spirited night attack that the 80th Regiment had made to silence a heavy gun that was harassing the British lines, and the trenches and guns were overrun against feeble resistance; some seventy guns were taken. But as Lal Singh's hard-tried soldiers streamed away in some disorder to the haven of the Sutlej, Tej Singh, with a large force of cavalry and nine-pounder guns, made a tardy appearance from the camp near Ferozepore.

In some desperation, the weary and depleted British formed a hollow square defensively about the village of Ferozeshah. However, Tej Singh, with either admirable circumspection or wary apprehension, stood off and began a steady bombardment of the British position. If Tej Singh had held to this plan to destroy the British at distance, Ferozeshah would probably have been a Sikh victory, for the British were too few and exhausted to attack and capture the tormenting guns. Then a mass of enemy cavalry approached the great British square and there, strangely, transfixed by some peculiar psychological influences, halted into inertia. Gough now ordered his own cavalry to charge and the splendid 3rd Light Dragoons, though much weakened by casualties, issued out, supported by a native cavalry brigade, to break up the dense enemy ranks and thrust them back in confusion. The late afternoon saw Tej Singh disengaging his forces and abandoning the field of Ferozeshah to the British. He moved off to Sobraon while the British, with some great relief, settled down to lick their considerable wounds.

Ferozeshah was a costly British victory, with 2,415 casualties, of which over half were suffered by the few Queen's regiments. Sir Henry Hardinge was to exclaim feelingly: 'Another such victory and we are undone!' Sir Harry Smith was a caustic critic of the muddled handling of the army, for if Littler can be blamed for rushing his weary division into action prematurely and unsupported, the whole general attack on the first day was badly co-ordinated; he described the army as one unwieldy battalion under one commanding officer who had not been granted the power of ubiquity. Michael Edwardes, in *A History of India*, is severe:

> 'The campaign of the First Sikh War does no credit to British arms. Officered by a Peninsular general of monumental stupidity and incredible bravery, the war was almost lost by a pathological belief in the virtue of the bayonet and a fantastic disregard for the efficacy of artillery.'

After Ferozeshah, the Commander-in-Chief clearly did have some second thoughts about artillery and its effects. He confided to a civilian of the Governor-General's staff that 'the fire was even more terrible than at Albuera, for the Sikhs had guns in position of treble the calibre ever used in European warfare'. Also, some concern and awareness arose in high places gradually about the suicidal role given to the 62nd in that first attack, particularly as Littler, seeking to explain the lamentable performance of his division, wrote a damning indictment, blaming the 62nd's repulse for the wretched showing of his sepoy regiments. Gough set out to redress the slur cast by Littler's report by issuing a general order to the Army of the Sutlej, part of which said:

> 'The chance of war placed the 62nd Regiment in opposition to the strongest part of the enemy's entrenched camp, at the battle of Ferozeshuhur, on the night of the 21st ult.
>
> 'The following day after the enemy's position had been carried, the Commander-in-Chief, with several officers of his personal Staff, minutely inspected the point attacked by the 62nd Regiment. He found it had been defended by numerous guns of heavy calibre.
>
> 'The very heavy fire by which the regiment was assailed and its steady, devoted gallantry under the storm are best attested by the fact, of which His Excellency was an eye

witness, of the space in front of and close to the enemy's battery having been thickly strewed with the bodies of the brave officers and soldiers who fell in the assault.

'The Commander-in-Chief finds that 18 officers and 281 men fell on this occasion, and that the regiment did not desist from its noble efforts to carry the position until ordered by the Brigadier commanding to fall back.'

On the same day of this order, January 28, 1846, 62nd reinforcements of four officers and fifty-one men under Captain William Mathias were in action at Sir Harry Smith's brisk victory at Aliwal. Marching in from Delhi on January 7, *en route* to the main army, they joined Smith's detached division south-west of Ludhiana and saw much fighting and strenuous work moving to the relief of Ludhiana. At the Battle of Aliwal, where Ranjur Singh's punitive force was broken with severe losses, they protected two vital eight-inch howitzers. On February 7, Sir Harry Smith's division and much-needed 400 reinforcements collected on the way rejoined Sir Hugh Gough and the army at Bootawallah, where it waited before the Sikh fortified camp at Sobraon for a siege train and its ammunition to arrive from Delhi. Matthias' detachment joined the 62nd, which had marched in from the Ferozepore on the 20th, on forward outpost duty at Rodawala, two and a half miles south of the Sikh positions. From Rodawala, Major Shortt, now recovered from his wound and again commanding, was summoned on the afternoon of February 9 to the Commander-in-Chief's tent where he found the Divisional Commanders and the Brigadiers also present. The siege train had arrived and the last act of the First Sikh War, the Battle of Sobraon, was about to be enacted.

Shortt's presence, amid this senior assemblage as a mere battalion commander, was doubtless due to the fact that the 62nd, in the battle plan for next day, had to carry out a preliminary and vital operation before the main attack on strongly fortified positions which the Sikhs, 30,000 strong, had sited on a bend of the Sutlej, with their back to the river, over which a pontoon bridge connected the right bank and Burra Sobraon.

> 'In the village of Chota Sobraon the enemy had posted a strong picquet within an entrenchment. It was necessary to drive in this picquet before Sir Hugh Gough could push forward his heavy guns within range of the great Sikh entrenchment, and when detachments of her Majesty's 62nd Foot stole cautiously down upon them in the darkness and mist of the morning, they were both found unoccupied, and were taken possession of without firing. It was afterwards ascertained that these posts were held during the day, and abandoned after dark in the evening; and this circumstance, added to a thick fog which deferred the dawn, was very favourable to the British, enabling the Commander-in-Chief to bring up his several divisions in order of battle, and post his artillery without any alarm to the enemy, in whose camp might plainly be heard the light song and rolling note of the nukaruh (kettledrum), which told of deep and false security.'

At 7 a.m., the British guns opened up, with the Sikh artillery coming into action to reply, and after two hours bombardment, the three British divisions crossed their startlines. On the left, Sir Robert Dick's 3rd Division, supported by the 1st Cavalry Brigade with the Cavalry Division's Headquarters, commanded by the one-armed Waterloo veteran, Sir Joseph Thackwell, was to assault against the right of the Sikh defences, which intelligence, in fact Sikh treachery, said was the weakest; the 3rd Division was strengthened for its role, with three brigades instead of the two the other divisions had, and the 62nd, now returned after its first light task, was brigaded with the 9th Foot and the 26th Bengal Native Infantry. Major-General Gilbert's 2nd Division and Sir Harry Smith's 1st Division, much weakened after Aliwal, were centre and right, each with a cavalry brigade. The grouping of the army, its rested condition and the preparatory bombardment with heavy guns were reassuring indications that Gough had learned some hard lessons at Ferozeshah.

The 3rd Division stormed over the Sikh right with great *elan* but once enmeshed inside, the forward battalions were soon in danger of being overwhelmed by masses of Sikhs who concentrated against them, almost ignoring the demonstrations of the other two British divisions. Ashburnham's brigade, with the 62nd, entered the entrenchements and struggled forward to join the bloody press with the bayonet. But the great weight of the Sikh numbers was still pushing the redcoated British and the sepoys back, until added impetus came to the attackers by Major-General Thackwell leading a squadron of the ubiquitous 3rd Light Dragoons in single file through the trenches and other obstacles. Later another squadron was followed by the 4th and 5th Native Light Cavalry. The cavalry made short charge after short charge and gave some valuable space for the infantry to reorganize and resume their attacks. The mounting strength of the cavalry also gave the 1st and 2nd Divisions an easier entry into the Sikh positions, despite the stronger natural and artificial obstacles on their fronts.

The Sikhs, now attacked at close quarters on three sides by a steadily growing enemy with bayonet and cavalry sword, began to break, though Joseph Thackwell was among several who said, 'many of the Sikhs were very brave and have defended themselves manfully.' The bodies of their fallen littered the ground thickly while their slowly-retreating hosts were pressed remorselessly back into the Sutlej, where the lifeline of the bridge of boats had broken under the pounding of the first flying body of horsemen. Hundreds drowned in the Sutlej, which had risen seven inches overnight from heavy rain, flooding the ford, and the Horse Artillery inflicted havoc and carnage on the mass of enemy struggling frantically across to the far bank.

By 11 a.m., the Battle of Sobraon was over, and with its British victory, the First Sikh War; five days later, the Sikhs sued for peace. British and sepoy losses were 3,500 while an estimate of Sikh casualties is 10,000. This time the 62nd fortunately was almost unscathed, with four killed and forty-three wounded.

The 62nd crossed the Sutlej with the main army and marched with it via Kasur to the Sikh capital, Lahore. When the peace treaty was ratified a month later at the Lahore Durbar, the Regiment returned to its peacetime cantonment at Ferozepore. In October, it marched back to Lahore when rebellion in Kashmir took troops from Lahore and Jullundur, but its Indian days were drawing to a close. On November 25, 1846, the 62nd marched out of Ferozepore for the last time, under orders to return to England.

This final, nostalgic march was through Delhi and Agra but at Allahabad, where command passed from Colonel Reed to Lieutenant-Colonel Shortt, the Regiment embarked on February 2, 1847, once again on boats on the Ganges. And, again, its waterborne course was dogged by ill-fortune. Another sudden Ganges storm dispersed and drove ashore many of the boats and, stranded, the Commanding Officer's craft caught fire; again, the Colours were lost, a mere four miles from the 1842 incident. The Regiment finally reached Calcutta on March 23 and soon afterwards began embarking on the *Adelaide*, the *Hindoostan* and the *Duke of Wellington*.

After seventeen years of tragedy, hardships and war in India, where so many of its officers, soldiers and families lay buried amid the tropical greenness of the Carnatic or the arid, semi-wastes of the Central Provinces and the Punjab, the 62nd Regiment assembled during the high summer of 1848 at Winchester amid the pastoral beauty of southern England.

Chapter 5

In the New Southland

In 1824, as the 62nd Regiment was settling down in the Irish establishment after Nova Scotia, a new regiment was born in Glasgow: the 99th. Almost sixty years later, both the 62nd and 99th were to come together in the Cardwell Reforms as the 1st and 2nd Battalions of the Duke of Edinburgh's Wiltshire Regiment.

The sixth and last regiment to be so numbered, the 99th was raised at the behest of the Governor of Mauritius, Major-General Gage John Hall. On March 25, the Regiment came into being under the command of Lieutenant-Colonel George Hardinge, with an authorized strength of thirty-two officers and 618 rank and file organized into eight companies. Its facings were deep yellow.

In December, 1824, the 99th marched to Edinburgh Castle and later, in January, embarked for Portsmouth where they remained until October, probably waiting for transports, before sailing to Mauritius. Twelve uneventful years were served in that picturesque island, which had been captured from the French in 1810, before the 99th returned to disembark in Cork in October, 1837. The equally pleasant routine of garrison life in Ireland, social military and sporting, was interrupted in 1841: the 99th were warned for convict escort duty to Australia. In September, the Regiment moved to Chatham where it was stationed with gradually diminishing strength, as each escort detachment went off, for almost three years before it re-assembled in the great, unknown south land.

Some 160,000 convicts were transported to Australia

Army Museums Ogilby Trust

A Shako of the 99th circa 1840.

before this barbarous system was stopped in 1868, and the need to guard these desperate men *en route* and at the convict settlements led to several infantry regiments being stationed in the Australian colonies. A young midshipman describes the scene on a convict ship off Chatham in those days:

> 'The remainder of the day was taken up in preparation for the first batch of convicts who were to arrive the following morning, embarking from the Dockyard. Early next morning before the Bosun and his mates had piped the crew to breakfast, the same Government craft came alongside with about 100 poor wretches under a strong guard. Boylike, I had been very anxious to see real convicts, but I remember being shocked at the first sight of them. They were clad in coarse yellow fustian stuff and broad check, spattered all over with the Crow's Foot or Broad Arrow, the Government mark. Each man had also coarse underwear and checked shirt, and they were closely cropped and shaven. They were shackled two and two and in their ill-fitting clothes, rough, low shoes and Glengarry bonnets presented a wretched and painful appearance.'

This miserable duty was not without tragedy for the 99th, as forty-six all ranks were lost at sea in two disasters, also with a number of their families. But as each ship discharged its wretched human cargo in Sydney Cove and the escorting soldiers rejoined their Regiment, life in early colonial New South Wales was far from unpleasant. The role of guarding convict working parties and fulfilling a presence for internal security was not arduous and the serenity and vastness of the bushland seemed to absorb and mutate the violent passions of these outcasts of European civilization. More and more convicts became emancipated and set out to hew a new life out of the virgin bush, as did their erstwhile gaolers, the time-expired soldiers, marines and warders.

However, if the only excitements to disturb the slow-moving colonial way of life on the Australian mainland was an occasional convict escape or a marauding band of black-fellows spearing stock on an outlying property, in the sister colony of New Zealand a serious rebellion was festering. There, the indigenous people, the Maoris, were of Polynesian stock, well-built and hardy, and much addicted to warfare, in which their tactics revolved about hand-to-hand fighting, as they had no missile weapons, and stoutly-constructed stockades, called *pahs*. Formerly pagans and cannibals, many had now been Christianized by missionaries over the years and but almost all debased by rum and attracted to muskets and gunpowder by contact with polyglot European adventurers who came to settle and trade.

One Maori chief who was incensed by the loss of anchorage money from whaling captains, assumed by the Crown in 1841, and the land encroachments of the *pakeha*, or white man, was Honi Heke, in North Island. In July, 1844, he rampaged with his tribesmen through the settlement of Kororareka and chopped down an offending flagpole flying the Union Jack. After a show of contrition, induced by the arrival of 170 men from the 96th Regiment from Australia, he again incorrigibly attacked the hated flagpole in January 1845. Finally, in concert with a renowned pagan chief, Kawiti, he attacked Kororareka with 600 tribesmen and drove the badly co-ordinated defence of a 96th detachment, seamen and marines to evacuation. The First Maori War was now on in earnest.

In April, 470 men of the 58th, 96th, marines and volunteers under Lieutenant-Colonel Hulme landed at the abandoned settlement from Auckland to dispose of the turbulent Honi Heke. The expedition was a dismal failure. Confronted, without artillery, by the formidable works of a *pah* defended by the natural valour of the Maori warrior fighting in his own elements, Hulme had to withdraw his sorry force back down over the rough mountain tracks.

To redress the wounds of this reverse, reinforcements were soon on their way from Sydney. In early June, the *British Sovereign* arrived in Auckland harbour with 200 officers and men of the 99th under their Commanding Officer, who was deputed 'to assume command of all troops in New Zealand with the temporary rank of Colonel on the Staff'. This choice of commander, Henry Despard of the 99th, could hardly have been more unfortunate; though an officer of some Indian experience, he was of a fiery and impatient disposition which was not alleviated by a generous share of bovine stupidity. Though he set about correcting the artillery deficiency from which Hulme suffered, his whole choleric outlook and limited military ability were unequal to the leading of a joint force in primitive terrain.

The assembly of the new expeditionary force at the Bay of Islands began on June 10, and six days later, the movement of the guns and heavy stores by boats up an inlet and the march inland began. Now the old painful struggle with the rude bush tracks, swamps and stony creek beds that had harassed Hulme set in, as the meagre bullock drays and horse-drawn carts toiled forward. At Waimate Mission Station, where the force rested five days, Despard met for the first time a redoubtable friendly chief, Tamati Waaka Nene, who had been the saving of the Hulme expedition and who had lately defeated Heke when the rebel attacked his *pah* to destroy him before the British took the field again. Waaka was to become a great favourite with the soldiers and was known as 'General Walker'.

At dawn June 23, Despard marched his force out from Waimate in the rain and cold of a New Zealand winter on the final trek to Heke's new *pah* at Ohaeawai; Major MacPherson, one of the original subalterns from the raising of the 99th in 1824, commanded 180 men of the 99th, and the rest of the force consisted of 270 men of the 58th, seventy men of the 96th, thirty Marines from H.M.S. *Hazard*, eighty Auckland Volunteers, and 250 friendly Maoris under Waaka. The going was so difficult that it was sunset before they covered the mere seven miles to Ohaeawai and musketry fire indicated that Waaka's scouts had driven in a rebel picquet from a conical hill overlooking the *pah*. The hill, which was to have key significance, was called 'Waaka's Hill'.

Ohaeawai *pah*, which had been greatly strengthened by Kawiti, now stood before Despard, though without Honi Heke, who was lying fourteen miles away with a bullet in his thigh from the fight with Waaka. Not especially large, it was built very solidly on rising ground. Each wall was blistered with projections that ensured enfilading fire along any section, and its timbers were young pururi trees thirty to fifty inches in circumference, sunk six feet in the ground with a height of ten feet exposed and laced together with vines and overhung with thick mattresses of green flax which absorbed musketry balls.

Next day, Despard set his four guns up and began bombarding, but the shells had little effect. Even when the battery position was moved up within 200 yards of the *pah*, the elastic and tenacious flax closed up about the shotholes and made it difficult to gauge the extent of any damage. On the night of the 25th, Despard decided to launch a concerted bombardment to force a breach and attack, but torrential rain cancelled the operation. Over the following days, the guns continued their desultory work, ineffective fire fights flared up between the rebels and their besiegers, another abortive attack was stood down and the watching and waiting was enlivened by two rebel sorties which were driven back. But by Monday, June 29, British casualties were slowly mounting: a grenadier of the 99th and a seaman were killed by enemy fire, and several were wounded.

Despard, ill-tempered and frustrated, now sent for a thirty-two-pounder from H.M.S. *Hazard* and settled down impatiently in the miserable wet and thundery weather to await its laborious arrival. By Sunday afternoon, he could bear the inaction no longer and determined to attack without the heavy gun. All his field officers without exception, including Hulme, were vehemently against such a suicidal plan. At this *moment critique* in Despard's blood pressure, news came that the vital gun had reached Waimate with twenty-five sailors from the *Hazard*. When it came up to Ohaeawai later in the day, the crowd of curious who flocked about it drew heavy rebel fire and another soldier was killed.

The whole of that night and the next day were occupied positioning the gun on which so much hope was now pinned. But as the sun shone in a rainless sky the following day, July 1, that hope was nearly dashed by a critical sortie from the rebel garrison.

The upshot of this brisk encounter was that the British flag from Waaka's Hill fell into rebel hands and later, Despard was mortified to see it flying in the *pah* upside down and at half-mast. This caused another upsurge of choler and he was now quite adamant on attacking without further procrastination. Under the orders of his acting Brigade Major, Lieutenant Rupert Deering of the 99th, his force assembled at 3 p.m. for the attack. The advanced storming party of two sergeants and twenty soldiers, all volunteers from the three regimental detachments, was led by Lieutenant Jack Beatty, of the 99th; then followed two assault parties, the first under Major MacPherson, 99th, with forty grenadiers of the 99th and forty of the 58th, and a second commanded by Major Bridge, with the sixty remaining grenadiers of the 58th and forty from the Light Company, 99th; a support group of 100 was under Lieutenant-Colonel Hulme, while a captain and forty soldiers of the 58th remained firm on the crucial Waaka's Hill; the balance of the force was under Despard's own hand.

Inevitably, the point of attack chosen by Despard, the north-west corner, was the strongest part of the *pah*. With a hearty cheer, the patient British infantry in their red tunics, still wearing the old-fashioned leather stock, all ragged and some barefoot, all carrying a full knapsack even in the assault, went forward on this forlorn hope. As William Free, who enlisted in the 58th in 1842 and died at the age of ninety-three at New Plymouth, New Zealand, in 1919, says: 'We just went at the strong stockade under orders from a Colonel who did not know his business and who had a contempt for the Maori.'

In ten minutes, under the horrified gaze of their commander, the slaughter of the attackers was fearful. The concentrated close range rebel musketry fire cut down forty fatally, among them Lieutenant Jack Beatty, and another eighty wounded, Major MacPherson seriously. In despair, Despard ordered the retreat to be sounded and the shattered withdrawing ranks, with some valour, recovered the wounded as they went. The bodies of the dead lay before the *pah* all that night which was rent with the shrieking and celebrations of the victorious rebels, causing the despondent soldiers to think that captured comrades, not yet dead, were being tortured. Next day the dead also lay where they fell, until the Reverend Burrows, from Waimate, arranged a truce which allowed the collection of the bodies the following morning.

As the dead were given a mass burial, it was a depressing and sobering time for the imperious Despard; his attack had been an appalling failure, his ammunition was running low and the morale of his men had been badly affected by the disaster. He now decided, as impetuously as he had rushed to the offensive, to abandon the siege and withdraw, but Waaka and other friendly chiefs, who with some discretion had refrained from the assault on the unbreached *pah*, protested vigorously and, after much argument, won a reluctant extension for the operation of two more days.

On July 8 and 9, haphazard firing broke out again, with the thirty-two-pounder opening up only occasionally due to lack of ammunition, but on the night of the 9th, a strange silence developed over the *pah*. In the early hours after midnight, Waaka's scouts, who had become suspicious because of the lack of sentinels' calls and then the sudden barking of dogs, crept up to the *pah* and after cautious investigation, found it to be deserted. The rebels had slipped away; blood had been spilt on the *pah* and according to one of the several curious Maori war customs, it was abandoned.

So, in spite of Despard's abject tactical handling, the operation against Ohaeawai ended somewhat fortuitously in an empty victory of sorts by virtue of occupation. Ohaeawai was destroyed, and also the *pah* of another rebel chief, Te Haratura, at Pakareka, on July 16. Despard and Hulme went back to Auckland to discuss future operations with the Governor.

Despard fretted at Auckland until he sailed back to the Bay of Islands on August 26 with reinforcements for both the 99th and 58th. There he was joined by another 214 men of the 58th from Sydney in early October.

The new objective crystallizing for the campaign was Kawiti's own *pah* of Ruapekapeka, or 'The Bat's Nest'. Kawiti had complained plaintively to Honi Heke that so far all the fighting had been on Heke's land; now he wanted some of his own local glory. On December 8, Despard moved out by river and track to the Bat's Nest with some 1,500 men; an Infantry Brigade of the two flank companies of the 99th, three companies of the 58th, eighty marines and some volunteers, and a naval brigade of 280 seamen of H.M.S. *Racehorse*, mainly manning the artillery. With them was the indefatigable Waaka Nene.

On the 27th, Despard and his field force stood before the Bat's Nest, sited on a narrow ridge with a deep and abrupt ravine between their camp and the *pah*. Despard turned the ravine by cutting a track through the brush and concentrated on the herculean labour of getting the guns up; as well as eight bullocks, sixty men had to toil to drag them up the rough and winding tracks. As this laborious preparation went on, a welcome reinforcement of a company of the 58th, despatched from the notorious penal settlement of Norfolk Island, arrived.

Over the next few days, Despard began advancing his guns methodically from position to position nearer the *pah* and by January 2, 1846, he had established his breaching battery of the two thirty-two-pounders in a clearing masked by fern and protected by a stockade wall in a wood a mere 400 yards from the *pah*. That afternoon, a strong force of rebels sallied out to attack the thirty-two-pounders but Waaka's men intercepted them and drove them off. The friendly Maoris celebrated this sharp little victory with a tremendous *haka* that shook the ground well within earshot of the *pah*.

On January 9, a recovered Honi Heke slipped into the Bat's Nest with sixty followers; he had heard that the British preparations for a final assault were now almost ready and again expected that the 'red tribe' would advance on to the muzzles of the rebels' muskets and be destroyed. The following day, a concerted bombardment roared out from the British guns, shrouding the *pah* wall in smoke and dust as the defenders took shelter in their dug-outs and even Honi Heke was shaken by its violence. This heavy concentration of fire began battering down the outer wall but behind, as daunting as ever, were the two other wall systems.

As the outer wall was well breached from the weight of fire, Despard now decided, despite the Ohaeawai debacle, to attack and a storming force of 200 was quickly mustered. But Waaka Nene and some of his chiefs ran over to him and protested bitterly against the folly of assaulting at this stage and, quite out of character, Despard reluctantly yielded. The gunners continued on with a harassing dropping fire but that night, another unusual silence, with no baiting calls from the rebel sentinels, enveloped the *pah*.

The next morning was a Sunday and in one of the extraordinary situations that happened in this First Maori War, it transpired that Honi Heke and most of the rebels, who were Christians, were outside the *pah* in the bush holding a service when Waaka's men crept up to the outer wall and deduced it was almost deserted. In response to their frantic silent signals, a number of soldiers and seamen ran up to join them and began moving into the *pah.* Inside was Kawiti and a mere handful of fellow pagans dozing blissfully in the trenches; surprised, they stood to their arms and fought their way out through a back entrance.

Meanwhile, the dutiful Heke and the other rebels, dramatically disturbed at their devotions by the sound of musketry fire coming from the direction of the *pah*, rushed back and found that they had to fight fiercely to try to recover their fortress. For some four hours, a battle raged, but the soldiers were too wily to be drawn into the bush after the rebels; some seamen did rashly leave the *pah* but were summarily cut down. Gradually, the dispirited rebels made one last charge, mainly to recover their dead and wounded, and then melted away into the bush in final retreat.

Again somewhat fortuitously, Despard was left the victor in the field and next day, he began demolishing Ruapekapeka, the 'Bat's Nest'. The British had lost twelve killed, mostly seamen in the foolhardy sally from the *pah*, and thirty wounded, of which the 99th fortunately had only one killed and one severely wounded. So ended the rebellion in the north, for soon afterwards, both Heke and Kawiti sued for peace.

Despard left 300 men of the 58th at the Bay of Islands and sailed back to Auckland with the remainder of his force. On January 26, he himself embarked to return to Sydney, 'much', as one caustic officer remarked, 'to the satisfaction of the troops in New Zealand'.

Meanwhile, Maori turbulence in the extreme south of North Island in the vicinity of Wellington was becoming serious. In late January, another company of the 99th joined the two flank companies at Auckland and to meet the new threat in the south, sailed in a force of 600 men under Lieutenant-Colonel Hulme for Wellington.

For the next eighteen months, the three companies of the 99th saw some excellent active service in the Hutt River valley area against a crafty and worthy enemy, Te Rangihaeata, whose warriors were well versed in minor tactics and the expert use of their native ground. After sustaining some light casualties in this guerilla phase, in August, 1847, the three companies left New Zealand and its Maori troubles to rejoin the Regiment in Sydney.

Next year, the 99th moved to Hobart, in Van Dieman's Land, soon to lose its title with such sinister convict connotations to be renamed Tasmania, and here it served its final eight years in Australia, with occasional company detachments in Melbourne and Norfolk Island. In 1856, the 99th Regiment sailed for the United Kingdom, leaving behind 400 N.C.O.s and soldiers who volunteered to serve in other Regiments so that they could take their discharge in the great, new land. Among their many unknown descendants in Australia, a monument today stands in Hobart in memory of the men of the 99th who fell in the First Maori War.

Chapter 6

Confrontation with the Tsarist and Celestial Empires

'62nd Regiment.—This regiment appears essentially in good order.'

When the pen of the Adjutant-General scratched this assessment at the Horse Guards in a letter dated August 29, 1853, the 62nd was again serving in the Irish establishment; after Winchester, it spent one year in North-West District and then crossed over to Ireland from Liverpool in 1849. Pointedly, soon after the Adjutant-General's report, it was placed at readiness to sail at short notice for Malta. Deteriorating relations with Russia were causing a build-up in that ever-sensitive British sphere of influence, the Mediterranean.

On February 2, 1854, three companies embarked at Queenstown, followed a week later by the other three, including the Grenadier and Light Companies. But, though war was declared on Russia in late March, the 62nd languished at Malta for some seven months until orders for its own move to the active theatre of operations, the Crimea, came. On November 13, the 62nd disembarked at Kamiesch Bay, a few days after the Battle of Inkerman.

As the Regiment marched, 566 strong, to the camp of the 2nd Division on Inkerman Heights, its arrival coincided with an atmosphere of deepening gloom for the British Army in the Crimea. The costly victories of the Alma and Inkerman had left units badly undermanned and the struggle for mere survival, with worsening weather and the appalling administration, was causing exhaustion and growing sickness rates. On the day after the 62nd's disembarkation, a catastrophic tornado devastated the rudimentary camps of the Allies and drove their ships aground; this disaster made the Army go on to half-rations for a prolonged period: half a pound of mouldy biscuit, half a pound of salt pork or beef and raw green coffee which had to be cooked with the few roots that could be foraged.

The 62nd was soon inducted into the misery of winter in the trenches before Sevastopol. However, through the ravages of an inordinately severe Crimean winter, operations by both the Allies and the Russians, either from the formidable garrison inside the city and the field army outside, for several months were at a low key. The close of the winter saw a conscience-stricken Britain pouring out money for huts, transport and medical supplies and equipment and the abysmal material plight of the British soldier began to improve. Early in April, the 62nd moved with the other units of the 2nd Division from Inkerman to the Woronzoff Ravine which ran from the city through the centre of the British front.

Two major objectives were now solidifying in the rather indecisive Allied councils: the Great Redan fort, with the broken ground of the Quarries before it, on the British front, and a kilometre to the north-east, the Malakoff Tower, with the hill called Mamelon in front of it, for the French.

It was not until early June, however, that the Allies were ready to assault their first phase objective. At 3 p.m., June 6, 600 Allied guns thundered out a powerful barrage until nightfall when heavy mortars took over; at 6.30 a.m. next morning, French infantry went forward against the Mamelon and two British columns, a mere 700 with another 600 in support, attacked the trenches and gravel pits of the Quarries.

Both preliminary objectives were cleared at bayonet point

National Army Museum

The storming of the Great Redan, Sevastopol, September, 1855.

and that night, the 62nd, which had been waiting in support, moved up to the captured ground to reinforce the assault troops and to organize the defence system so that it now faced the Redan. Those hours of darkness were to be packed with desperate, savage action, where the defenders fought with entrenching tools, rocks and fists when their ammunition gave out, as several Russian counter-attacks erupted against them; a young Captain Garnet Wolseley of the 90th, later the famous Field-Marshal, was always to recall that night as the most exhausting of a long campaigning life, and he was one of the many who collapsed insensate with tiredness among the dead after the last enemy attack had been beaten off at dawn. Among the 62nd's casualties was the Commanding Officer, Robert Shearman, an officer who had distinguished himself nine years ago at Sobraon; also two company commanders were killed, and Captain William Ingall, a casualty at Ferozeshah, again wounded; command of the Regiment devolved on Captain James Daubeny. Total enemy casualties were estimated at 5,000 killed and wounded, the British only 670, but the French, who in their impetuosity had carried on wildly towards the Malakoff, lost 5,500.

Ten days later, the Allies were ready to go for the main objectives, the Malakoff and the Great Redan. After a day-long bombardment from 600 siege guns and from the Fleet, the Allied assault infantry moved up during the night to their forward assembly areas. But it was a brilliantly clear midsummer night and the Allied preparations were easily detected from the city, where, by 2 a.m., the forward defences were jammed with Russian infantry and guns, waiting for the attack. When the French assault went in before dawn, it failed with terrible slaughter. The watching British Commander-in-Chief, Lord Raglan, as first light came, felt he must participate in the attack in the higher interests of Allied solidarity, despite a growing realization of imminent tragedy, and so the British assault troops, from the 4th and Light Divisions, emerged from the Quarries to cross 500 yards of open ground covered with long rank grass and pitted with shellholes to the Great Redan.

As the unsilenced Russian guns blasted out canister and roundshot and every embrasure on the Redan blazed a storm of fire, the attack was shot to pieces. One hundred yards out from the Redan was an abatis, five feet high and four feet wide, and here many survivors sheltered, deaf to the entreaties of their officers who quickly fell riddled with bullets as they stood up to exhort their men forward. In despair, the retirement was ordered and the remnants ran back through the smoke to their old trenches. Fortunately, the 62nd emerged unscathed from this debacle, as their role was only in support.

As the routine of trench warfare and artillery exchanges resumed, Lieutenant-Colonel Tyler arrived to take command of the 62nd from Brevet-Major James Daubeny. In July, the French, who had been receiving a steady stream of reinforcements, numbered 120,000 whereas the British maintained a mere 50,000 and it was their continuing weakness, undermined by daily losses in the trenches, that precluded an early return to the offensive. But by September 8, the Allies were ready to assault again.

For this operation, the assault troops were again drawn from the 2nd and Light Divisions, who occupied the trenches in the vicinity and who knew the ground and the enemy only too well; unfortunately, their ranks had been recently filled with young, raw reinforcements whose only knowledge of action was from the comparative safety of their fire trench, and this was particularly so regarding the much-used Light Division; in retrospect, it is clear that the rested 3rd Division and Highland Brigade should have been employed. The storming element of the 2nd Division was found equally by the 41st and the 62nd.

During September 5 to 7, the usual severe Allied bom-

National Army Museum

After the battle: the interior of a gun position in the Great Redan, Sevastopol, September, 1855, depicting the fury of the bombardment and the ensuing close quarter struggle.

bardment rained down on battered Sevastopol and on September 8, the Allied attack went in at noon. Within half an hour, the French tricolour was floating bravely over the Malakoff and the Little Redan but on the British front, after gallant penetration for a tenuous period, bitter failure set in. Many of the shaken young soldiers braved the fire storm to gain the abatis or the ditch about the Redan but once there, could not be budged. A heroic 100 stormers, many of them 62nd, did scramble up the rubble about the ruined face of the great bastion and entered it to hold it against ever-swelling odds for a noble hour. William Russell, *The Times* correspondent, wrote:

'The 62nd Regiment went into action 245 all ranks. . . .

> Lieutenant Blakeston was shot getting through an embrasure in the Redan. Lieutenant Davenport was shot through the nose. On the parapet two officers were killed or died of their wounds, and four officers wounded out of a total of eleven; three sergeants were killed and four wounded out of a total of sixteen; one drummer killed out of eight; fourteen rank and file killed and seventy-five wounded. . . .'

Among the fallen was again the 62nd's Commanding Officer, Tyler, who died of his wounds six weeks later.

The sheer weight of massed Russian infantry drove the gallant 100 back through the embrasures of the Redan and as the leaderless mob of young soldiers in the ditch were bombed with grenades and riddled with musket balls from the jeering enemy above them, the survivors of the stricken attacking force fled brokenly back to the haven of the Quarries. The French too had been ejected from the Little Redan but held violently to the Malakoff, despite the threat of mines whose wires were coolly found and severed.

As the bloodied and dejected British brooded in a sombre mood of disgrace, some consolation came that evening when observers reported a mass exodus of Russian troops pouring across a recently-built bridge of rafts to the other side of the harbour. Though the British had suffered in barely three hours roughly the same number of casualties as the eight hours of Inkerman, 2,450, and the French 6,000, the Russian commander, Prince Gortschakoff, had appreciated that his long-tried troops and the battered city had absorbed enough. The pounding of the Allied guns was inexorably reducing Sevastopol and his own losses had been heavy, with his very basic medical cover unable to cope. After a night of explosions rocking the ruined city, as the arsenal and dockyards were blown up, the Allies entered the abandoned suburbs the next morning.

The Crimean War was now in effect over. In November, James Daubeny, who had again assumed command of the 62nd, was promoted to Brevet-Lieutenant-Colonel and the

Regiment soldiered uneventfully in the Crimea until the following summer.

Peace officially came in April, 1856, and the following month, the 62nd embarked at Balaclava. Some twenty-four all ranks were mentioned in despatches for gallantry in the final attack of September 8 on the Great Redan, more than any other regiment, and later it received a prominent share in the distribution of French, Turkish and Sardinian decorations. The recipients of the awards from the Emperor of the French depict how much the old British Army owed to Ireland: Sergeant William Reilly, Drummer Thomas Finnigan, Privates Thomas Carney, Thomas Brophey, James McCarthy and James McKee.

The 62nd sailed away from the Crimea, not to return to the Mother Country after its eighteen months of active service but in the somewhat heartless military expediency of the day, to another overseas station: on June 2, 1856, the Regiment landed at familiar Halifax, Nova Scotia, and marched into Citadel Barracks. William Ingall, recovered from his wound, now arrived to assume command from James Daubeny, who remained as Second-in-Command.

When the American Civil War broke out in 1861, the 62nd were moved over to Canada as part of a precautionary build-up of British troops, and by train and sleigh in the severe snowstorms of the Canadian winter, occupied stations at New Brunswick, Kingston and Quebec. Finally, on July 6, 1864, the Regiment embarked on the *Himalaya* for England and after an excellent crossing, landed at Portsmouth to entrain for Aldershot to rejoin the 2nd Brigade.

After fifteen years abroad, the 62nd returned to an Army in England where reforms brought about by the administrative chaos of the Crimea were now gradually taking effect: the establishment of a proper system of supply and transport, a much-needed revolution in soldiers' feeding, instigated by M. Soyer of Soyer Stove immortality, even a Staff

College to train professional staff officers rather than having the old custom of a collection of decorative aristocratic amateurs on a General's staff, and some thought and care given to the ordinary soldier's well-being—what the modern Army calls man-management.

Perhaps that terrible winter of 1854–55 and the bloody assault on the Great Redan had not been in vain after all.

As the Crimean War ended, the 99th sailed from Australia for Ireland. A year in the Curragh, with much local recruiting, amply replaced the 400 soldiers who remained in Australia, and it moved to Aldershot in December, 1857, over 1,000 strong. From the rude society of the Australian colonies, the Regiment now progressed, with the superb flexibility of the British Army, to a far more rarified environment. As Colonel Kenrick relates:

> 'The 99th were extremely prominent during their nine months in Aldershot. Very smart in drill and dress, they were always chosen to find the Guard on the Royal Pavilion, and were in consequence nicknamed "The Queen's Pets". Also their officers wore more than the usual amount of gold lace on their uniforms. The 99th's sartorial perfection at this time is said to have given rise to the expression "dressed up to the Nines", as other Regiments in Aldershot were constantly trying to achieve the same standard.'

The 99th's Aldershot domination of that focal point of military Victoriana, the parade ground, ended in December, 1858, when, as a result of the turbulence caused by the Indian Mutiny, it sailed from Portsmouth for Calcutta.

As garrison of Fort William, the 99th, like the 62nd twenty years previously, found the sultry, sprawling city, with its temptations and debilitating climate to affect fresh-cheeked young soldiers, an unhealthy station that caused some 100 fatalities from disease during the fourteen months sojourn there. It was doubtless with some relief, as well as the natural exhilaration of the prospect of active service, that the Regiment received orders to join an expeditionary force assembling at Hong Kong for operations in a fascinating new sub-continent of British mercantile and political penetration, China.

The 99th, under Lieutenant-Colonel George Reeves, disembarked at Hong Kong in March, 1860, and went into camp at Kowloon on the mainland. Here the expeditionary force of some 14,000 strong, organized into two infantry divisions, a cavalry brigade, and two batteries of Armstrong guns, was concentrating under a spare, lean Scots cavalryman, Lieutenant-General Sir James Hope Grant.

The background to the coming conflict was prevarication by the Court of Peking in ratifying the Treaty of Tientsin of 1858, which allowed for diplomatic representation, trading concessions and missionary activity. An expedition mounted impatiently by the British in 1859 to force the issue had foundered ignominiously in the mud flats by the Taku Forts, guarding the mouth of the Pei-Ho river to Peking, where Admiral Sir James Hope had disastrously landed a force of seamen and marines, and lost 434 killed and wounded and four gunboats. The French were to be deeply concerned through the barbarous death of a missionary and they were also assembling a punitive force at Shanghai under Lieutenant-General Cousin de Montauban. However, unlike the Crimea where their administrative preparations were far superior to the British, they were much delayed in their readiness by having to procure cavalry horses in Japan and make other *ad hoc* arrangements. The British brought their horses from India and solved much of their administrative difficulty by enlisting several thousand Cantonese coolies who, dressed in a semblance of a uniform and with a band like a naval rating's ship's tally about their wicker coolie hat proclaiming 'Military Train', were cheerful and hard-working rogues whose main drawback was to prove that they were almost uncontrollable in looting and raping their northern cousins, which they did with incredible aplomb.

In April, as the 30-day ultimatum to Peking expired, the British force sailed north and landed bloodlessly at the rather useless stepping stone of Chusan. Here the 99th, now under James Day, spent two pleasant waiting months until June while the British sailed further north again to a forward concentration area at Talienwan Bay for the invasion of the mainland. Finally, on July 26, in a formidable spectacle the British and French fleets, with their transports, made a landfall off the mouth of the Pei-Ho and the defended Taku forts guarding the way to Peking and Tientsin. Two days later, the British and French, avoiding the formidable mud flats and the guns of Taku, came ashore at the

National Army Museum

The captured Taku Fort, during the China War of 1860, with dead Chinese soldiers in the foreground and the French and British flags flying over one of the redoubts. The British flag was planted by Captain John Hart Dunne, of the 99th (later General Sir John Hart Dunne, K.C.B., Colonel of the Regiment 1898-1924).

abandoned village of Pei-tangho, ten miles to the north of the forts.

However, the dreaded mud of the Pei-Ho was not entirely avoided and the British landing was led by a bow-legged little brigadier, barefooted and without his trousers which were slung over his shoulder, swearing volubly as he led his brigade in single file and similarly informally clad, over a mile of smelly, tidal mud. Once ashore, during a ten-day build-up period, the national characteristics of the two Allies asserted themselves; the British built roads and a jetty while they noted sourly that the French looted and hunted for food.

The diligent Hope now advanced on the forts, carefully building bridges over the many canals as he went forward to give himself openings for manœuvre, and taking the two fortified villages of Tank-ku and Sin-ho. By August 21, the attack on the forts was ready to go in. Hope persuaded Montauban with some difficulty that the easternmost of the two on the northern bank was the entire key to the whole siting of the three forts, one being on the southern bank, and the main effort was directed at it. Despite the heavy British guns, it was a formidable task for infantry to assault, with both wet and dry ditches, abatis and a belt of pointed bamboo stakes about it, and if the abortive expedition of 1859 was any pointer, the Chinese fought their guns well and bravely. The enemy commander was a Mongol general named Sang-ko-lin-chin and his name was promptly regularized by the British troops to 'Sam Collinson'; later a splendid soldiers' rumour went around the camps that he was in fact an Irish deserter from the Royal Marines.

The British artillery was too heavy for the fort and at dawn next morning loud explosions were heard from each of the northern forts. The Allies launched their infantry attack and after three and a half hours fighting, clearing a bravely resisting garrison with the bayonet, the eastern fort fell. The 99th, to their disappointment, had been left at Pei-Tangho to guard the vital base but Captain John Hart Dunne, with the initiative of a keen young officer, had obtained a staff attachment for the advance and was soon in the captured fort to plant the Union Jack on its upper structure. The other two forts submitted almost at once and 'Sam Collinson' rode hurriedly off with an escort of 150 horsemen towards Tientsin.

Success followed rapidly on this sharp little victory. The Viceroy of Chihli surrendered not only Taku but Tientsin, and by August 26, the Allies were in the great river port. Word now came that the Imperial Commissioners were arriving from the Court, shaken by the Taku reverse, to resume peace talks and a feeling arose that the war was over. However, a bombastic Imperial decree came from the young and debauched Emperor:

> 'Last year, the barbarians tried to force the entrance of the Peigo but within a moment's time their boats were sunk, and thousands of enemy corpses were floating on the waters, as far as a mile from the shore.
>
> 'I had hoped that such a lesson would make the barbarians more circumspect, but this is not so; less than a year after their defeat, they have returned in greater numbers and more insolent. Taking advantage of the low tide, they have landed at P'ei-tang and have attacked the forts of Taku. But being barbarians, they attacked them from behind. Our soldiers are used to meeting the enemy head on, and never expected such cowardice and treachery.
>
> 'The barbarians have occupied Tientsin, and are proud of their success, instead of being filled with shame. My wrath will soon overtake them, and exterminate them without pity. I hereby command all my subjects, Chinese and Manchu, to hunt them down like evil animals. Let every village that these devils may be nearing, be evacuated. Let all food they might capture be destroyed. Their cursed race will thus be overcome by hunger, and will perish like fish in a dried-out pond. Given in Yuan-min-yuan, on the 23rd day of the second moon of the ninth year of our reign.'

The 99th now moved on with the main body, acting as

advance troops, to Hosiwu, where the Allied Army halted while Lord Elgin and Baron Gros, the Allied political heads of the expedition, sent their envoys to meet the Imperial emissaries at Tungchow, only twenty miles from Peking. Far from being starved, the soldiers of the 99th and the other 'barbarians' settled down in this waiting period to enjoy the bounty of the North China countryside, and the local merchants and peasants, once they realized the red-coated barbarians meant no harm, flocked to sell their fat sheep and oxen to the Commissariat and the men peaches, pears and apples and even blocks of ice to cool their drinks.

Treachery at Tungchow, which bore the Allied envoys away to Peking as captives, set the Allied Army in motion with some determination and on September 18, it came up against an Imperial army of 20,000 under 'Sam Collinson' at the town of Changkiawan. Hope Grant was not in too balanced a situation, with Napier's division and part of the French at Tientsin, from whence they were being urgently summoned. However, he attacked, with the French on the right and the 99th forming part of the British centre. A serious turning movement by a mass of Tartar cavalry around the British left flank was broken up by the Armstrong guns and exploited by the British and Indian cavalry, and with a thrust forward by the 99th and the other infantry in the centre, the Chinese retired in confusion through the town and retreated.

From information gleaned by his scouting parties and local inhabitants, Hope Grant and Montauban learned that the defeated army, with considerable reinforcements, had taken up a new position about the splendid marble and stone bridge of Palichao over the Yang-liang canal which bore fresh water from the Pei-Ho to Peking.

The Allied battle plan was that de Montauban, with his 5,000 infantry, supported by artillery, would go for the vital bridge of Palichao while the British, with less infantry but more cavalry, would move on the left flank to cross by a wooden bridge and drive the Chinese right on to its centre. While the two European contingents advanced with disciplined cohesion, it was also clear that the Imperial army, too, was behaving with set-piece coolness. The Chinese main arm was its Tartar cavalry which trotted forward in well-controlled squadrons, and its artillery was to the fore, with the infantry entrenched as a solid pivot. But the whole Chinese battle line was working to a central authority; rather like a fleet at sea, orders were transmitted by waving banners. The central banner waver was a huge Tartar standing on the Palichao bridge who bore a great yellow and black banner, and all subordinate commanders kept watching him intently.

The Tartar cavalry began to move forward in two great segments: one moving into the gap between the British and French which the French succeeded in breaking up with concentrated rifle fire and by bringing up their artillery, and the other began the customary out-flanking movement about the British left. However, this was shattered by the charge of the Kings' Dragoon Guards, Fane's Horse and Probyn's Horse, and followed by the 99th and a battalion of Royal Marines with three guns, this mixed force poured over the wooden bridge and turned in against the Chinese camps and entrenchments. A final, disordered stand was made by the Chinese in front of their main camps but this disintegrated under the concentrated rifle fire of the 99th, advancing forward in skirmishing order and halting to discharge withering volleys. The few brave Chinese soldiers whe stood to the last were bayoneted and the 99th moved through their camps, burning as they went; the cavalry pursued to within six miles of Peking. Meanwhile, the French had carried the Palichao bridge with great *èlan* and were debouching over the canal to join the British.

The army rested at Tungchow where Napier's division, the small siege train and French reinforcements came up, and on October 5, Hope Grant advanced rather cautiously through old fortifications to the Imperial capital. He need not have worried as the weak and degenerate Court was paralysed with fear at the approach of the 'barbarians' and on October 8, succeeded in stopping the Allied operations with more vague promises of peace talks and the return of their eminent prisoners. As the Anglo-French force settled about Peking, the British suddenly lost the French who went off happily to sack the Emperor's Summer Palace five miles north-west of the city. Not even the rigidly disciplined British, held in check by summary application of the lash, could resist this splendid edifice crammed with priceless jade, silks and precious stones, and they followed. However, the sensible Hope Grant ordered that all the British 'liberated' objects should be auctioned, with one-third of the proceeds going to the officers and two-thirds to the men; from this, a soldier received £4 and all were content.

National Army Museum

The final objective of the 99th in the China War of 1860: the Anting Gate, Peking.

Grant's plan to take the city was to force an entry by the Anting Gate, on the north-east sector of the walls, and deploy in the broad streets of the Manchu city. However, with his guns laid ready as his ultimatum to surrender expired at noon on October 12, the Gate was surrendered, the Emperor and his lackeys fleeing in panic to Jehol, and the Allies entered triumphantly and peacefully. On October 24, the Allies staged a ceremonial entry, for Lord Elgin and Baron Gros to sign the Peace Treaty, and the 99th's Colours were carried in Lord Elgin's Guard of Honour.

The weather was now changing perceptibly and the Allies were determined to return south by November 1 before the bleak North China winter set in. The great city of Peking was spared but the beautiful Summer Palace and 200 surrounding buildings were razed to the ground with fire before the victorious British and French marched away.

The 99th sailed south to Canton, and after a year there, spent another three years in Hong Kong. One company was detached to Shanghai to assist in the protection of British interests against the depredations of the Tai' Ping rebels, a major revolutionary movement of curious Protestant origin but which later became blatantly anti-foreign, and which was a major thorn in the side of the Manchu dynasty.

In early 1865, the 99th sailed away from Chinese waters, but until the demise of the Wiltshire Regiment, some of its best pieces of silver came from the ill-fated Emperor's Summer Palace. To give complete respectability to this onslaught of redcoated and redfaced barbarians on the heart of the Middle Kingdom, Captain Hart Dunne presented the Queen with a Pekinese 'Lion Dog' which he had found wandering in the grounds of the Palace. This first Pekinese, aptly, was named 'Lootie'.

Chapter 7

Clash in Africa

AFTER the Crimean War and the sideshow of the China War of 1860, somnolent Victorian peace set in, disturbed only by the minor wars of India. Long-awaited military reforms slowly appeared: the daily pay of a private in Line regiment was raised from 13d (taken by 8½d for his food, 3½d barrack damages and the upkeep of his uniform, and 1d beer money) to 15d in 1867, and the tunic replaced the coatee; after 1870, the spiked helmet of the much admired Prussian infantry appeared. To eradicate the abuses of billeting, much overdue barrack building ensued. In the important field of infantry weapons, by 1869 the muzzle-loading Enfield rifles had been changed to the breech-loading design of Mr. Snider, and already rumour spoke of the introduction of the new hammerless Martini Henry rifle.

The 62nd (Wiltshire) Regiment spent these halcyon years in Ireland and Aldershot until 1869, and then Indian service followed, its stations sounding like the beat on some grand Imperial drum: Lucknow, Dum Dum and Delhi. In 1880, it sailed for the molten rocks of Aden, then a distant responsibility of India.

Meanwhile, the 99th (Lanarkshire) Regiment had left China in 1865 for service in South Africa. Four years in Natal and Cape Colony presaged a later close and active service link with the vast sub-continent. For, ten years later, after being stationed in Scotland and Ireland, the 99th disembarked again at Durban in 1879 to join the campaign against the warrior Zulu nation.

A group of officers of the 99th in splendid variations of dress at Cape Town, 1869. The Kilmarnock forage cap with the diced red and white border denotes the origin as a Lowland regiment.

In 1879, the fratricidal and bloodthirsty Zulu military monarchy, with its standing army of 40,000 organized into regiments called *impis*, was considered to be threatening Natal, Transvaal and Swaziland and intriguing with other tribes. When Cetewayo, the Zulu King, rejected an ultimatum to disband his army, on January 12, 1879, 6,000 British troops under Lord Chelmsford, with their native auxiliaries, invaded Zululand in three columns: left, or northern, thrust under Colonel Evelyn Wood, V.C. (gained as a midshipman before the Great Redan in the Crimea!), centre, and from the west, the main column via Rorke's Drift and Isandhlwana under the overall commander, Chelmsford, and right, or southern, moving north over the

Tugela river along the hills by the Indian ocean, under Colonel Pearson. The objective of these three widely dispersed forces was the royal kraal at Ulundi.

Two companies of the 99th were left at Durban and Stangar as local garrisons while the main body of six companies marched north to join Colonel Pearson's command on the Lower Tugela. On January 12, Pearson crossed the river into Zululand with his column split into two groups, of which four companies of the 99th under Lieutenant-Colonel William Welman moved in the forward body. Though Zulu scouts were seen moving along the ridge lines as the British advanced steadily through the bushy country, it was not until the 22nd that a brisk action developed along the Inyezane stream, the attacking Zulu force being driven off with only minor losses to the British of two officers and eight soldiers killed, and sixteen wounded. Next day, Pearson reached the mission station at Etshowe where he had the intermediate objective of establishing a forward supply base for the final advance on Ulundi.

However, January 22 had been a fateful day for British arms elsewhere than on the Inyezane. At Isandhlwana, a Zulu army had overrun the camp of the central column, killing 858 officers and men, and, chastened by this disaster, Chelmsford proposed to Pearson on January 28 that he should evacuate his exposed forward base and retire back to the Tugela. This suggestion, however, was somewhat academic by the time it reached Pearson on February 11, for on that day the Zulus, who had been circling Etshowe since January 24, severed the track back to the haven of the Tugela, where half the force still lay. If anything, oblivious of the Isandhlwana defeat, Pearson had weakened his strength by sending back two companies of the 99th and two companies of the Buffs to escort forty-eight empty supply wagons.

Etshowe was to remain in a state of siege until early April. But it was a very gentlemanly affair, with the Zulu

A. Bernard Mills, Durban

The Zulu War, 1879: the fort and mission station of Etschowe.

not pressing close, and whilst the garrison improved the defences with some urgency, the Bands of the 99th and the Buffs played daily. On March 1, a punitive expedition of four companies of the Buffs, one of the 99th, some sappers and Marines set out to destroy a kraal seven miles away and to improve the food supply, for now the garrison was committed to eating its own very tough trek oxen. On March 20, a runner came through from the Tugela to warn the garrison that a relief column was starting out on the 29th.

During February and March, heavy rains had flooded the Buffalo and Tugela rivers and the consequent breakdown in movement is thought to explain the strange Zulu inaction after Isandhlwana. It was in torrential rain on the appointed day that the relieving force for Etshowe, under Chelmsford himself including five companies of the 99th as the Durban and Stangar companies had rejoined, crossed the Tugela.

At noon on April 1, Chelmsford set up an entrenched camp by the Ginginhlovo stream, in an open area clear of scrub and long grass. Large groups of Zulus had been seen and during that night, many campfires flickered in the nearby hills. At dawn, Chelmsford sent out his mounted infantry to scout and by 6 a.m. he had reports that a Zulu army was approaching.

In the two enveloping horns that overflowed about three

Illustrated London News

Rear-guard of the 99th Regiment crossing the Amatikulu River

sides of the British *laager*, the brave Zulus attacked but though the full-blooded impetus of their assault carried them within a mere twenty yards of the trenches, their assegais and few firearms were futile against the withering fire of breechloading rifles and two deadly new Gatling guns. The violence of their attack was savagely halted and as they wavered in decimated ranks, the British horsemen moved out to cut off their retreat, followed by the howling pack of the native contingent. In one and a half hours, and suffering few losses, the British inflicted 1,200 Zulu casualties; and on April 3, Etshowe was relieved, the whole force moving back to the Tugela.

A period of some reorganization now set in, as other Regiments arrived from England and overseas stations as a result of the reverse at Isandhlwana. The 99th was in the 1st Brigade of 1st Division on the Lower Tugela while the 2nd Brigade remained forward at Ginginhlovo. This low-lying country on the coast was unhealthy for the British and enteric fever was rife; also the recruitment of native drivers and transport oxen for the final advance against the Zulus was proving very difficult. It was not until mid-June that 1st Division moved north along the coast, burning a few kraals *en route* and capturing some prisoners and cattle. But the advance, which was supplied over the booming surf beaches along the Indian Ocean by the Navy, was fairly pedestrian and dull, and all the glory was to be won in the centre and north; on July 4, Chelmsford broke the Zulu military power at the Battle of Ulundi and though Sir Garnet Wolseley rode into the camp of the 1st Division as the new C.-in-C., the war was already over except for the hunting down of Zulu stragglers. Cetewayo their leader was captured on August 18.

After a parade on July 23 at the operational beach port of

Illustrated London News

Graves of men of the 99th Regiment at Fort Pearson

Port Dunford, 1st Division disbanded and soon afterwards the 99th moved back into Natal where its headquarters were established at Pietermaritzburg until its embarkation for Bermuda on December 30. So ended a minor but colourful campaign that was the epitome of military Victoriana: hard marching at the end of an ever-lengthening and tenuous administrative backing struggling to cope with undeveloped country and where the superior discipline and fire control of outnumbered Line regiments triumphed over a brave and barbaric foe.

The 99th had barely a year in lovely Bermuda before orders came to re-embark in March, 1881, for Natal. The First Boer War had broken out and the British had sustained defeat at Majuba Hill on February 27. The 99th landed at Cape Town in early April and deployed in Cape Colony but palsied political direction in Whitehall had decided to make peace. On December 31, the 99th sailed away again, to India.

That same year, 1881, the 62nd embarked in the torrid heat of Aden, after a hard year, for England.

Meanwhile, since 1871, when Edward Cardwell became Minister of War under Gladstone, a wide and sweeping re-organization of the Army had begun. As early as 1873, the 62nd Wiltshires and 99th Lanarkshires were linked in a gradual system of evolution whereby Infantry Regiments of the Line with only one battalion would eventually develop into a two-battalion Regiment, with one battalion overseas and the other at Home, and a regimental depot in their county of recruitment. In 1874, the title of the 99th was changed to 'The Duke of Edinburgh's Regiment', and it was given permission to bear the Cypher and Coronet of the Duke as its Regimental badge. It still maintained its association as a Lowland Regiment, and wore a diced border on its shako and forage cap until 1886.

Accordingly, a Regimental Depot was set up in Le Marchant Barracks, Devizes, Wiltshire, in 1878 and the

Imperial high noon: the 2nd Battalion at Mandalay, Burma, 1895.

depot companies of the 62nd and 99th moved in with additional training staff for that important adjunct of the Regular Army, the Militia. Finally, in 1881, the process of evolution ended with somewhat brutal finality: the time-honoured and glory-covered numbers of Regiments were abolished for territorial titles, to a great outcry, in vain, from most of the Infantry. Official amalgamation came about, and, *inter alia*, the 62nd (Wiltshire) Regiment and the 99th (The Duke of Edinburgh's) Regiment became the 1st and 2nd Battalions, The Duke of Edinburgh's Wiltshire Regiment. The facings of the new Regiment were buff.

After Aden, the 1st Battalion was stationed in England and Ireland until 1895, when it embarked for India. The 2nd Battalion soldiered in India, after its abortive return to South Africa, at the cantonments of Rawalpindi, Pershawar, Nowshera, Cherat, Ambala, Amritsar and Jhansi, and finally, Fort Dufferin at Mandalay, Burma, until returning to garrison the forts about Portsmouth, also in 1895.

However, the 2nd Battalion's belated home service was soon to be disrupted: In October 1899, the Second Boer War erupted, and on December 16, the Battalion embarked at Southampton on S.S. *Gascon* for the theatre of war. A fortnight after disembarking at Port Elizabeth on January 12, 1900, it was in action among the plains and kopjes of northern Cape Colony.

The 2nd Wiltshires sailed as the Black Week disasters of Stormberg, Magersfontein and Colenso were reverberating about a shocked powerful and proud Britain. After landing, they were hurried north to the Colesberg area where Major-General French was holding a wide and tenuous front against an Orange Free State invasion force of 7,000 under Koos De la Rey and Piet de Wet. On January 25, they were in action in a bold, aggressive plan by French to seize the

The composite Volunteer Company from the two Wiltshire Volunteer Battalions parading at Le Marchant Barracks, Devizes, in 1900 before going out to South Africa to join the 2nd Battalion as J Company.

important Plessispoort defile through the hills to cut off the enemy invading force from its communications across the Modder River.

With his two cavalry regiments riding deeply to the rear, French advanced his infantry along the ridgeline to the pass. By 2 p.m. this flanking movement had gone well, reaching a favourable position within 1,500 yards of the suspected Boer positions, and from this startline, the 2nd Wiltshires were sent forward. Under light enemy fire, six lines of Wiltshiremen, 100 yards apart, marched steadily forward towards the objective. The local commander then sent them to ground while he asked French for permission to press home the attack in force. But the experienced French was worried; the cavalry were late appearing, the afternoon was lengthening and he was deeply suspicious that the rocky slopes were crammed with concealed Boers holding their fire. At 4 p.m., yielding to his premonitions, he ordered retirement, and as the Wiltshire firing lines rose to obey, the frustrated Boers opened up with a blaze of artillery and Mauser rifle fire.

> 'But in spite of the heavy fusillade which followed them back, the Wiltshires retired as steadily as they had advanced and rejoined the column with the loss of but 10 men wounded.'

As Boer pressure mounted in early February, the Wiltshires were not so unscathed. French departed with the cavalry on February 6 to lead the Cavalry Division in the van of the main army Lord Roberts and Kitchener were assembling secretly to strike at the hearts of the two Boer republics, and 12 Brigade in which the 2nd Wiltshires served, was left with the unenviable task of containing the enemy at Colesberg, covering the two key railway junctions of Naaupoort and De Aar, and by remaining there, preventing the local Boer districts from rising to join the Free Staters. But now the energetic De la Rey had been reinforced to some 11,000 with more artillery and between February 6 and 12 began enveloping the flanks of the main British position at Slingersfontein. The Wiltshires were holding the left flank and after fierce close fighting at Hobkirk's Farm, from which an embattled half company was rescued by a dashing counter-attack by a squadron of the Victoria Mounted Rifles, retracted to a tight defensive position as the Boer assault mounted in strength.

With his left forced and his right equally assailed, the Brigade Commander decided to withdraw to Arundel. Orders were issued for all troops and movable stores to concentrate during the night February 13/14, with the baggage to move off to Arundel at 3 a.m. and the marching troops at 5 a.m. As the situation worsened the previous evening, the orders were changed to accelerate the withdrawal to begin at midnight. However, as the weary Brigade moved into Arundel at dawn, it was discovered that two companies of the Wiltshires, deployed on outpost duty, were missing. Through a muddle at battalion headquarters, their commander, Major F. R. Macmullen, had not been informed of the changed timing for the withdrawal.

Macmullen retired his 150 men on the original timing but now he moved through a countryside alive with hostile probing forces, and he was soon attacked and trapped. Though he was reinforced by a few mounted troops and RHA guns sent to locate and rescue the lost Wiltshires, Macmullen was to fall mortally wounded among twelve dead and forty-four wounded, with the remainder of his men overrun and taken prisoner. War historian Conan Doyle is one authority who absolves Lieutenant-Colonel Henry Carter from this calamity:

> 'No man in that trying time worked harder than Colonel Carter of the Wiltshires (the night of the withdrawal was the sixth he had spent without sleep).'

But despite the forceful De la Rey driving the Boers

advancing on Arundel into another attack, the British beat off this waning attempt and the whole situation was suddenly eased by many of the Free State burghers hurrying back north to Bloemfontein, for Kitchener was now pounding A. P. Cronje in a bloody battle at Paardeburg, on the banks of the Modder, *en route* to the Orange Free State capital. Gradually, the initiative passed back again, as reinforcements were received while the apprehensive enemy thinned out, and by February 28, the force was back at Colesberg, including the battle-scarred Wiltshires. On March 13, Roberts entered Bloemfontein with the main army and two days later, the brigade crossed the Modder to join him.

For the next two months, the 2nd Wiltshires stayed in the Bloemfontein area, suffering a bad onslaught of enteric fever. As Roberts resumed his march on Pretoria and the Transvaal, 12 Brigade moved to the rugged eastern region of the Orange Free State to be one of the four strong columns attempting to pen in the major surviving strength of this battered Boer republic, 9,000 fighting men under President Steyn and tried leaders as Christian de Wet, Olivier and Prinsloo, against the mountains along the border with Basutoland. 12 Brigade, based on Senekal, moved against de Wet at Bethlehem. On July 7, the 1st Royal Irish, supported by four companies of the Wiltshires, seized the high ground on the enemy's left flank and by afternoon, the Boers were in full retreat to the south to hold five main passes, or *neks*, in the mountains.

On July 23, in a violent downpour of rain and snow, the 2nd Wiltshires and the other battalions of 12th Brigade came up to one of the more formidable of these passes, Slabbert's Nek.

> 'The mighty clefts in the mountain barrier looked impregnable. Through one the road twisted at right angles, appearing to end in a solid wall of rock, while cliffs, sometimes 2,000 feet high, rose up on either side to merge into precipitous plateaux and ridges which in one direction

> presently fell away to a great drop to the other pass, similarly flanked by forbidding heights whose crevices and caves gave the Boers perfect cover. They were well-prepared with artillery and they manned the lower slopes as well as trenches across the passes.'

The sheer strength of Slabbert's Nek forbade a frontal attack but while the Munster Fusiliers held the centre with the massed British artillery pounding the Boer entrenchments, two companies of the Royal Irish and another two of the Wiltshire worked their way up a spur that ran down from the enemy left. But darkness found the Wiltshire's checked, with some forty-two casualties. Then, in the cloak of night, a patrol reached a dominating height that the Boers had failed to picquet and by 4 a.m., it was firmly reinforced by four companies of the Royal Irish and two of the Wiltshires, commanded by the appropriately named Colonel Guinness of the Royal Irish. During the remaining hours of darkness, the force moved perilously along a goat track following the crestline, with men crawling on hands and knees at times with a sheer drop below, but first light brought the exhilarating reward of the Boer positions spread before them below. Their strong position turned, the Boers abandoned Slabbert's Nek.

Clements followed up and a stiff action developed at Slaapkranz. At the head of three of his companies which captured a key farmhouse position on the enemy left flank, Lieutenant-Colonel Henry Carter was severely wounded twice and Captain Bolton, assuming command, held on throughout a hard day until reinforcements came up and the Boers evacuated that night. With the end of July, the Boer general Marthinus Prinsloo and 4,500 burghers surrendered to General Sir Archibald Hunter and his four columns, and operations in the eastern Orange Free State ceased. The 2nd Wiltshires then left 12 Brigade and after an excursion of two months back in the Arundel area when commandos made another foray into northern Cape Colony,

'A weary job, 20 miles a day the average, and sometimes we had to do over 30.'

they spent the remaining year of the war in the arid bush-veld area of north Transvaal, based on Pietersburg, seeking out and attacking, and defending against, in that final guerrilla phase of the war one of the last active Boer leaders, C. J. Beyers.

When peace was signed at Vereeniging on May 31, 1902, the 2nd Wiltshires had lost thirty-five fatal battle casualties and eighty-five wounded, some 107 were accounted missing, and another sixty-nine died of disease. Never in the great tragic defeats or major bloody victories, they were a steady Line regiment which trekked endlessly over a seemingly inexhaustible veldt after a mobile pony-mounted enemy who was an excellent shot and who knew how to use ground and cover. If glory eluded them, steadiness and devotion to their duty never did.

Chapter 8

The Great War

ON THE lovely summer's evening of July 27, 1914, the officers of the 1st Battalion at Tidworth were about to set out to dine with the 4th Battalion, the Regiment's unit of the newly-created Territorial Army then under canvas at nearby Bulford, when a cycle orderly arrived at the Mess to say that the Commanding Officer or Adjutant were required on the telephone by Headquarters, Southern Command. The prosaic appearance of the cycle orderly presaged a train of events to mobilization and war. On August 13, after much hectic activity, the 1st Battalion, was crossing from Southampton to France with the 3rd Division.

Only the year before, the 1st Battalion had returned home after eighteen years abroad; from 1895 to 1909, it had soldiered in India, at the stations of Karachi, Peshawar Cherat, Rawalpindi and Jhansi, and from 1909, in South Africa, at Pietermaritzburg. The 2nd Battalion, after the Boer War, had come back to England in 1903, and after serving at Pembroke Dock, in Ireland and at Gosport, was now in Gibraltar. One month after mobilization, the 2nd was sailing back to join the newly-forming 7th Division.

From the tens of thousands of enthusiastic volunteers who streamed into the recruiting offices in that dying summer of a dying world, three Service battalions of the Wiltshires were quickly raised at Tidworth: the 5th, 6th and 7th. The 4th Battalion sailed in September with the Wessex Division for India to release Regular units, while the 3rd (Militia) Battalion became a training and holding unit for the Regi-

1914: the Band of the 7th Wiltshires, on a recruiting tour of the villages of north Wiltshire, marching through Sherston. It is interesting to recall that the well-known writer and poet of the Great War, Siegfried Sassoon, took the name of George Sherston for his hero in Memoirs of a Fox-hunting Man, *and when the 1st Wiltshires were Demonstration Battalion at the School of Infantry, Warminster, 1953-55, he lived alongside their camp at Heytesbury House.*

ment in the South of England for the duration of the war.

One of the First Hundred Thousand was Private Tom Moor, destined for the 2nd Battalion, whose reminiscences of those stirring days have a decided Dickensian flavour:

> 'There were times to be merry, and times to be sad. I had worked for United Dairies 1902 to 1914 in the making of cream cheese. I started at 3/6 a week and finished at 22/6 in 1914. That was a princely sum in those days. Twelve years service and still a kid. The future looked good to me. Devizes, my birthplace, was a military town with Barracks where they trained men for the Colours as Sam Darling trained Derby winners at Beckhampton close by. I little thought at that time they would be giving me the "once over". There it was. Then came 1914 and the war. I can see the Wiltshires marching off to Southampton, then France, all time-serving men fit as fiddles. . . . I still have the bob Major Stewart gave me. Fancy, a bob a day and a

twenty-four hour day at that. Mine was a "tanner", as I allotted my mother the rest.'

Ten days after leaving England, the 1st Battalion was in action with the B.E.F. at Mons. The retreat to the Marne, with stands at Solesmes and Le Cateau, followed in those long, hot summer days. Again the discipline and fire control of long service British regulars overcame the confusion and exhaustion of the great withdrawal before the German hosts.

After the strategic victory of the Marne, the 1st Battalion was with the B.E.F.'s advance to the Aisne. Early in October, the B.E.F. left the Aisne for the north in the famous 'Race to the Sea', and here the 2nd Battalion appeared with the 7th Division, after the landing at Zeebrugge on October 2. While the 1st became heavily committed about Neuve Chapelle, the 2nd were soon swept into the bloody First Battle of Ypres in late October and November, the last phase of open warfare before both adversaries, worn and

Imperial War Museum

Men of the 1st Wiltshires returning from an attack bearing captured trophies.

with heavy casualties, settled down to trench warfare. Christmas 1914 was to see another last touch of the old world so rapidly disintegrating. Tom Moor describes a unique battlefield scene:

> 'All through Christmas Eve the Germans are playing mouth organs and melodians. The war is not worrying them. Nor us. We sing "Christmas Awake." All night we keep it up, one against the other. We are waiting for the dawn and a double issue of rum, if we are lucky. As the mist rises, there they are sitting on the tops of their trenches. Neither side fires . . . we climb warily out and through the wire. . . . We edge up closer while the officers talk it over. Rum war, we think. Then we see it's the Saxons! But the savagery of war is still grimly present. We wandered around. The ground was as hard as a bullet. There were a good many Warwicks lying there. It made your blood boil. We could not bring them back but we would bury them, and Jerry lent us a hand.'

1915 saw the slogging matches of static trench warfare develop, with the 2nd Battalion taking part in the brilliant minor victory of Neuve Chapelle on March 10, and the 1st in the Second Battle of Ypres, starting on April 22, where the Germans used poison gas for the first time. As the spring battles gave way to the summer, the Service battalions of the Regiment were soon on the move to active service after their month of training. On June 30, the 5th Battalion sailed with the 13th Division to Alexandria and, ultimately, the Gallipoli Peninsula, and on July 20, the 6th landed in France with the 19th Division.

The 5th were soon in action on the Peninsula when the 13th Division landed at Cape Helles on July 18 to relieve the hard-tried Regular 29th Division. After a brief respite on Mudros, the 5th went ashore with their Division at Anzac Cove in readiness for the great offensive on the night August 6/7 against the dominating ridge of Sari Bair while the newly arrived IX Corps landed just north of Suvla Bay as the other arm of the pincer. By August 10, the 5th

Imperial War Museum

A young subaltern of the Wiltshires who has been wounded in conference with a brother officer.

Battalion and the 6th North Lancashires were astride vital high ground when both units were overwhelmed by a violent Turkish counter-attack of divisional strength led personally by Mustapha Kemal; the commanding officer was killed and half the battalion was lost. The shattered 5th were moved to Suvla Bay where the remnants were in trenches before Scimitar Hill when the great freak storm of snow and floods struck the Peninsula in November. Suvla Bay was evacuated on December 20 but after another brief rest at Mudros, the 5th's brigade was sent back to Cape Helles to cover the final evacuation there, which it did successfully until the last night, January 8, and returned to Egypt.

In France, the 6th Battalion was nursed along in trench reliefs at Laventie and Festubert. But a great battle loomed ahead in September, which did not open auspiciously:

> 'On the 21st, orders were received for the forthcoming Battle of Loos. On the evening of the 24th, we moved up into the reserve trenches, and on the 25th, at 5.15 a.m., received orders that a gas attack followed by an infantry attack would take place. Our Brigade was to attack the German trench by Rue d'Ouvent as soon as the Division at Givenchy signalled that they had been successful. Unfortunately, though they were at the start successful, they were forced back again and when the 9th Welsh and 9th RWF, supported by the 9th Cheshires and 6th Wiltshires, advanced, they were badly enfiladed and suffered very heavy casualties. The 6th Wiltshires fortunately only advanced one company, "D", but even then had 40 killed and 75 wounded.'

In September came the 7th Battalion's turn to cross to France but their appearance on the Western Front was to be very brief. Two months later, the 7th sailed from Marseilles for Greece where an Anglo-French force had landed at Salonika to give support to the Serbian Army reeling under massive attack from Austrian, German and Bulgarian armies. For timid political reasons, the British

Imperial War Museum

Officers and men of the 1st Wiltshires after their return from the fighting at Thiepval on the Somme in September 1916.

Imperial War Museum

A Company, 7th Wiltshires, being inspected at Salonika, 1916.

division was forbidden to push north towards the Serbians as the French did, and the 7th found itself digging and wiring rather futilely in a defensive position about Salonika.

On the Western Front, 1916 bore the 1st, 2nd and 6th Battalions to the great carnage of the attrition Battle of the Somme, which dragged to a muddled and exhausted negative close in mid-November with terrible casualties. The 6th, who sustained 380 killed, wounded and missing on the Somme, also fought in the battles on the Ancre from October 22 to November 29.

Meanwhile, in Egypt, barely a month after their arrival from Gallipoli, hard news fell on the 5th Battalion. Up to strength with 500 reinforcements and revived by recreation in Alexandria and hard training in the desert, they received orders to embark on February 14 for the Persian Gulf and the Mesopotamian Relief Force. In a violent thunderstorm on March 4, the Battalion disembarked at Basra and after a fortnight's fitness training, moved up the Tigris in flat bottomed boats that made two knots against the five knot

Imperial War Museum

The 1st Wiltshires go over the top, Thiepval, 1916.

current. At the end of March, the 5th were near the front and Lieutenant-Colonel R. I. Scorer, M.C., then a platoon commander, relates the approach to action:

> 'A day at Orah and we crossed the river, relieving the 7th Division in the Aum-el-Hannah trenches. A day or two spent in trench improvement, making of dumps, perfecting of jumping-off places and we were ready for the Turk. We went over the top at dawn on April 5th and drove him out of three lines of trenches, advancing to Falahiyeh where we were held up about noon and suffered a fair number of casualties. I remember on the morning of the 5th it was frightfully cold and we were promised a rum ration.'

Unfortunately, to the rage of the whole company, it transpired that the company cook had drunk the lot.

> 'The Falahiyeh position was captured that night and the Turks retired to Sannaiyat.'

Diarist Lance Corporal S. W. Blythman of 'D' Company is more terse but perhaps well depicts the atmosphere of this isolated campaign in a land of temperature extremes, of floods, marsh and desert, and primitive communications. He goes on to describe the 13th Division's unsuccessful attack on the Sannaiyat position, a strongly-fortified Turkish main defensive system surrounded by open, marshy ground barring the advance on Kut; earlier, the 7th Division had failed to carry it and was now dug in 600 yards away.

> 'Attacked at dawn on April 5th, capturing Turkish positions and advancing about 2½ miles. Got cut up in broad daylight—not a stitch of cover. Rested in the captured positions for three days. Moved into advance trenches April 8th and made attack at dawn on Sannaiyat position. Was wounded in the shoulder and back, managed to crawl in at night very weak from loss of blood. (It was a slaughter. I heard later that they had an 8 hour armistice to clear up the battlefield.) I was placed in a mulecart and taken to 19 Indian Field Ambulance. On April 11th, we were

put on a barge to go down-river, but the steamer could not shift us off a mud bank, so we were left there without food and water for 24 hours, and we were all stretcher cases.'

Another major attack by 7th Division foundered against the strength and marshes of Sannaiyat, but the splendid Russian victory at Erzerum, in north-eastern Turkey, suddenly caused the Turks to abandon their forward Mesopotamian positions to release troops to stem the Russian incursion, to the great relief of the baffled British. The 5th Battalion, 1,100 strong on landing at Basra, now went out of the line a mere 300. The summer came on, with temperatures up to 120 and 130 in the shade, and tropical diseases abounded. For the rest of the war, the 5th served with distinction during much hard fighting in the advance to Baghdad and beyond to Kirkuk.

In the same broad theatre, the Middle East, the 1/4th Wiltshires (an offshoot, the 2/4th, remained at Delhi) belatedly appeared on active service after its long sojourn in India when it joined 75th Division in Palestine in September, 1917. Under Lieutenant-Colonel Alan Armstrong, 'the Little Colonel' who was to show a great penchant for personal patrolling, the Battalion was blooded a month later among the wadis and cactus gardens of Gaza in the successful Third Battle. Arduous fighting in the Judean hills during the advance on Jerusalem followed, and after the capture of Jerusalem on December 11, the 1/4th became engaged in close operations between Jerusalem and Rafat in early 1918.

In September, 1918, the 1/4th moved with their Division to the coastal plain in the north-west for Allenby's master plan to break open the tenuous Turkish front with his infantry divisions and unleash his superb British-Anzac-Indian cavalry on the enemy rear. The last entry in their C.O.'s diary is on September 12 when his battalion is in the divisional rest camp and he is commenting on the success of the battalion sports meeting and an officers' dinner night, to which his G.O.C. came and at which his amateur battalion band played with great aplomb. One week later he was dead, killed in action on September 19 leading his battalion in the attack on El Tireh. This was the last battle the 1/4th Wiltshires fought, as Turkish resistance collapsed under the great breakthrough and peace soon followed.

In the strange, politics-ridden campaign in Macedonia, at last, in April, 1916, came a concerted move to advance the Allied force from its barbed wire cage about Salonika to the Serbo-Bulgarian frontier. The British corps, with the 7th Battalion, moved up the valley of the Vardar River, with its right flank resting on the Struma, through magnificent scenery that was also forbidding to military operations. The Bulgarians duly invaded in force in August but were held, and a checkmate type of war in these mountain fastnesses of limited attacks, raids and patrolling developed. In April 1917, as a diversion to a main offensive mounted by the French near Monastir, the British attacked west of Lake Doiran on the night of April 24.

> 'The ground was a wild jumble of hills divided by ravines and strongly-entrenched by the Bulgarians; in fact, the "Grand Couronne" and its offshoots formed one of the half dozen strongest field fortresses of the war.'

The 7th Battalion, with its brigade, attacked the Grand Couronne but the whole British assault failed against fierce defensive fire and deep wire, and the 7th suffered heavily, with over 300 casualties.

The main offensive was also a failure and the campaign sank back into its twilight torpor, with the malarial mosquito a far more formidable foe than the Bulgarians. The 7th operated in the Lake Doiran area in trench warfare for their remaining months in Macedonia.

In France, during 1917 and 1918, the holocaust and

ruthless decimation of modern siege warfare conducted with the vast material resources of the twentieth century continued. In 1917, the 2nd Battalion was in the Battle of Arras April 9-11 and both the 1st and 6th were at the Battle of Messines in June. All three Wiltshire battalions were embroiled variously after the Third Battle of Ypres erupted on July 31 until it ended in November, sucked fast in the deep mud of the infamous Passchendaele ridge.

A first-rate platoon sergeant of those days in the 2nd Battalion, Sergeant T. E. Brewer, who had enlisted as a band boy in 1905 and later became Regimental Sergeant-Major in the Twenties, and, finally, a distinguished Major (Quartermaster) with forty-four years service in the Regiment, describes a typical battalion attack over the top in the Western Front in a smuggled letter home:

> 'Imagine us marching up to our positions in the front line on a cold and wet night. Just as we were getting into position, Fritz lets fly with his guns but we got very few casualties. Luckily, we got into position and waited for the appointed hour to attack, which was dawn. This period is always the worst, not that we fear anything, but there is a certain nervous tension, wondering what is going to happen. After going with my officer to see that the men were comfortable, I sat down to rest a little, but could not. . . . The minutes crept on, the order came to get into No Man's Land. I followed my officer and we got the lads out into the open and laid down. This was very dangerous, as at any minute we might be detected and that would spoil the whole show. Fritz seemed to be uneasy, as he sent up a number of lights and his guns were firing spasmodically. I looked at my watch, ten minutes to go, another whisper to the officer about things in general, five minutes. I was not nervous now, and the men were calm. Suddenly, before one realized it, our guns opened up with their full force. It seemed as if hell was let loose; shells, lights, signal rockets all in one blaze. Fritz got his guns going but we were clear as he strafed the trenches we had just left, but our reserves caught this. Our barrage lifted from the German front line, we were off! Hardly conscious of anything, we were dashing for Fritz's front line. With

a cheer, we swept over on to his second line. We dealt with a few in his front line, and now we began to get some trouble. His machine guns and snipers began to tell. Some of my boys went down, but we could not stop to look in the dim light of the dawn. I saw my officer in front, we followed close on his heels. A bullet whistled close by my head. By this time we were in his second line and Fritz was running like a hare and we mowed them down. But we had a third and final objective to take, so forward again. On the way to the third line we came across a big concrete dug-out full of Huns. We called on them to surrender, they would not come out, so in went half a dozen bombs. Most of them were killed, however we left one man to deal with any that came up, but only one badly wounded man ventured out. We reached our final objective and started to consolidate and make a position. About 10 a.m. it started to rain and by night we were up to our knees in mud and wet through. Fritz peppered us with his guns, but the boys hung on. We remained up there for four days until at last we were relieved, and we waded through the mud a distance of six miles into comparative safety, but Johnny followed us all the way down with his shells. At last, just about dawn, we reached our bivouacs tired, wet and plastered in mud, where there was some nice rabbit stew and tea, also a tot of rum, which all went down grand. Then we crept into our bivouacs and slept, the first real sleep since the scrap started. Fritz still shelled about 500 yards away but they were only stray ones and we were too tired to trouble much.'

All three Wiltshire battalions were assailed in the first great Ludendorff offensive in 1918, when under the fortuitous cloak of a thick mist on March 21, some thirty-five German divisions drove deeply into the weak Fifth Army and part of the Third on the Somme. The 1st Battalion, now with the 25th Division since late 1915, had only joined the Third Army the previous December and was in reserve in the Bapaume area. A bitter defence ensued over five confused and bloody days, culminating in an ordered withdrawal, and the tenor of the resistance offered to the waves of field grey assault troops may be indicated by the award of the Victoria

Lt.-Col. R. J. F. Hayward, V.C., M.C., as a subaltern company commander in France, seated in centre. With him, left to right, are other officers of the 1st Battalion: 2nd Lt. Clarkson, 2nd Lt. Brown, 2nd Lt. Jones, Artists Rifles (later killed in action) and 2nd Lt. Bidwell.

Cross to Captain R. J. F. Hayward, M.C., an original 6th Battalion officer then with the 1st:

> 'For the most conspicuous bravery in action, this officer, while in command of a company, displayed almost superhuman powers of endurance and consistent courage of the rarest nature. In spite of the fact that he was buried, wounded in the hand, and rendered deaf on the first day of operations, and had his arm shattered two days later, he refused to leave his men (even though he received a third serious injury to his hand) until he collapsed from sheer physical exhaustion.
>
> 'Throughout the whole of this period the enemy were attacking his Company front without cessation, but Captain Hayward continued to move across the open from one trench to another with absolute disregard of his

> own personal safety, concentrating entirely on re-organizing his defences and encouraging his men.
>
> 'It was almost entirely due to the magnificent example of ceaseless energy of this Officer that many most determined attacks upon his portion of the trench system failed entirely.'

Only three officers and fifty-four other ranks reassembled in rear when they were finally extricated. In the incredible way of those gory days, the shattered 1st quickly absorbed a massive induction of reinforcements and moved well to the north back to the Ypres area. Here the Battalion had the supreme misfortune of deploying in the path of the second German offensive, along the Lys on April 9, and the whole carnage occurred over again; one of the many missing was the Commanding Officer, Lieutenant-Colonel S. S. Ogilvie, a triple D.S.O., later found to be a prisoner. Again, more huge drafts and the Battalion was back in action on the Messines ridge about Kemmel towards the end of the month, where the newly-arrived Commanding Officer was killed in action. Finally, in this desperate phase of the war, the 1st moved south again with the 25th Division to the Aisne to help hold the third enemy offensive there on May 27.

The same dramatic chain of events also enveloped the 6th Battalion, near the 1st at Bapaume. A note in the Battalion War Diary by the Commanding Officer, Lieutenant-Colonel Lord Alexander Thynne, who was wounded, tells all about an exhausting week of valiant defence and fighting withdrawal:

> 'The stand made by the Battalion at Morchies from 4 p.m. on the 22nd March to 5 p.m. on the 23rd, and all that it meant, is a glorious episode in the history of the Wiltshire Regiment. It is a heroic record of self-sacrifice stemming the victorious rush of a superior enemy and a model lesson of a rearguard fight. The subsequent retirement from Fremicourt through Bapaume to Grevillers and thence to Bayencourt was only one endless and stubborn

fight. Suffice it to say that only 1 officer, 1 sergeant and 18 other ranks came out of the struggle, and that officer had also been wounded.'

The 6th too went north afterwards, to become embroiled in the offensive on the Lys after taking in 550 reinforcements, including a new Commanding Officer who was to fall in the Lys battle. In May came the tragic order to break up the remnants of the 6th, except for a cadre, and the survivors were posted to the other two battalions.

The 2nd Battalion also had only returned to the Somme in January and was occupying a forward trench system opposite enemy-held St. Quentin when the full early fury of the great attack hit it on March 21. By the afternoon, the battalion was overwhelmed, having been surrounded with the enemy pouring by its flanks, and some twenty-two officers and 600 men were lost. Again the pitiful remnants were built up with reinforcements and fought also against the next two enemy offensives on the Lys and the Aisne.

As the German juggernaut expended itself in five convulsive offensives between March and July, the Allies moved slowly but surely back to recapture the strategic and tactical initiatives, and the three veteran Western Front battalions of the Regiment fought on, reconstituted, throughout the remaining climacteric days of 1918. The 7th Battalion came to France from the great valleys and panoramas of northern Greece in July and was also in the final Allied offensives, in action on the Somme and at Le Cateau, and finally, across the Sambre. Most happily, from its cadre, the 6th was resurrected in June, only a month after its demise, and took part in the final attacks and advance, as fitted its origin among the First Hundred Thousand, that ended in decisive Allied victory.

When the thousands of hot, smoking guns along the long, tortured front were stilled finally on November 11, 1918, so ended a terrible and great conflict in which it is impossible to chronicle and depict the strain and privations of prolonged trench operations, the appalling casualties, and the countless acts of heroism and sheer guts that happened day after day in raids and patrols and the normal train of operations.

Disbandment came in 1919 for the 5th and 6th Battalions at Devizes, in France for the 7th, and the 1/4th at Trowbridge; the 2/4th came back from India and disbanded in 1920. The Regular 1st and 2nd Battalions, after the initial great rush of demobilization, slowly rebuilt back to full Regular content and stood by for posting for the demands of Imperial policing in the worldwide Empire. Meanwhile, some 200 Regular veterans, sailed off to Archangel in May, 1919, with the 2nd Hampshires as part of the North Russia Relief Force.

The memories of the Western Front, of Ypres and the Somme, and of Gallipoli and Palestine, of Macedonia and Mesopotamia, were to be annually brought vividly back to life over the years with staunch gatherings of Old Comrades of the Regiment at reunions in the cities and towns of the West Country, London and Birmingham. A County regiment had given liberally of its blood, though now from staunch men drawn from all over England, in a holocaust where the artillery shell and machine-gun were the grim arbiters. The survivors, with their memories of those terrible years, could never forget.

Chapter 9

1939-45

'WHEN the news that war had been declared reached Bangalore in India on 3rd September, 1939, four young officers, who were sitting together in a bungalow, rose and stood in silence for a moment. Before three of them met again six years later, two had commanded battalions in action and had been wounded, and one had led a company in battle. The fourth was killed in Normandy at the head of his men.'

These young officers, Captain N. C. E. Kenrick, Lieutenants H. M. A. Hunter, L. H. Wood and R. M. C. Thomas, were then serving in the 1st Battalion which had moved to India from Malaya in 1936. During twenty years of imperial policing, the Battalion had served in Dublin during the Irish troubles 1920-22, at Tidworth and Plymouth until 1929, then Egypt 1930-32, followed by ten months in gay and turbulent Shanghai until it moved to Malaya in late 1932.

In 1920, the title of the Regiment was changed to 'The Wiltshire Regiment (Duke of Edinburgh's)'.

The 2nd Battalion, in 1939, was stationed at Catterick, in 13th Infantry Brigade, 5th Division. Its peregrinations over a globe solidly coloured an imperial red had been similar; in 1919, it disembarked in Hong Kong for three years, with one company detached to old familiar ground of the 99th's, Peking, as Legation guards, then India, at the stations of Kamptee and Secunderabad 1922-29, followed by a year among the hectic distractions of Shanghai. A sober, strict and highly efficient existence followed at Plymouth under a new and memorable Commanding Officer, later to be General Sir William Platt, the victor of Keren, 1941, and after a move to Aldershot, the 2nd sailed in September 1936 to Palestine and the Arab rebellion.

In this world war, it was the 2nd Battalion who crossed to France a fortnight after the outbreak. Well up to strength with reservists, the 2nd landed in France on September 14, 1939. The first months were spent constructing defences along the River Lys, so ominous from 1918, but on May 2, the Wiltshires were relieved and moved to Amiens under warning to go to Norway. This was decisively cancelled by the German invasion of the Low Countries on May 10.

Five days later, the 2nd were under artillery and air action in a defensive position at Hal, where their Brigade had deployed along the Charleroi canal to cover the withdrawal of the B.E.F. from the Louvain area. When the forward troops had passed through, a series of rearguard positions followed north of Enghien and Herrines. Finally, 5th Division was ordered to hold the line of the River Scarpe while the only limited counter-thrust conducted by the B.E.F. was delivered south of Arras by 50th Division and 1st Tank Brigade. On March 20, the 2nd Wiltshires marched into their allotted sector.

> 'The river valley here resembles the Wylye Valley between Warminster and Salisbury. To the south rises the high ground around Monchy, which overlooked the whole battle area, and was occupied by the Germans. The Battalion frontage was some 3,500 yards, and the Commanding Officer found it necessary to have all four companies up, with the Carrier Platoon in reserve mobile. On the right "D" Company held the Fampoux bridge area, and in the centre "B" Company held the Rouex bridge area. "A" Company occupied the gap between these two area companies. On the left, "C" Company held the quarry area, joining with the 2nd Battalion The Royal Inniskilling Fusiliers on our left.'

After the limited success of 50th Division's attack, heavy

shelling of the 2nd Wiltshires' area continued through the night of May 21 and the next morning. The next night a patrol reported no sign of the enemy across the river, and next morning, to get a bridgehead for the Cameronians to occupy the dominating high ground at Monchy, 'B' Company crossed the river. Once across, the company came under heavy machine-gun fire from concealed positions, and finally only thirty men could be extricated under artillery fire. This serious loss left a gap in the battalion locality which the Carrier Platoon and 'C' Company tried to fill.

That afternoon a major enemy assault built up, with the familiar curtain raiser of low-level bombing attacks followed by heavy artillery bombardment. By 9 p.m., enemy assault troops had forced a crossing and were infiltrating in force about the large area the Wiltshires were striving to hold, and twenty minutes later, the order to withdraw came. In confused, bitter fighting all companies fought their way out, though 'C' Company especially lost heavily. About midnight, the battered Wiltshires collected at Fresnes, two miles away, covered by the Carrier Platoon, and a mere 300 strong marched back north of Douai under further Brigade orders.

So ensued the exhausting ritual of a retreat under enemy supremacy and severe harassment. After a stand on the Ypres-Commines canal between May 25 and 28, on the night of the 28th the battalion disengaged to march over miles of crowded and confused roads to its M.T., and the final move to the beaches of Dunkirk began.

Finally, with 13th Brigade now down to one composite battalion, its Wiltshire element 261 strong, embarkation came in the early hours of June 1 from the mole at St. Malo. The 2nd Wiltshires still had fight in them:

> 'Three weeks of retreat and withdrawal and confusion had not taken the fighting spirit out of the Battalion, and the order that we could not embark with Brens was received with dismay.'

As the weary, depleted 2nd Battalion disembarked on its native soil, the 4th and 5th Battalions were standing to in East Anglia, both in 129th Brigade, 43rd Wessex Division. In this war, the 5th were a Territorial battalion, formed from the great expansion of the T.A. in early 1939, while the 6th, whose war record in the Great War had been so splendid, was relegated to be a Home Defence battalion throughout; similarly, the 7th Battalion, until disbanded in 1944 to reinforce the active battalions of the Regiment.

While the 1st Battalion languished in India, and the two Territorial battalions went through the hectic round of invasion scares and strenuous exercises to prepare for the day of battle, the 2nd Battalion was soon again on the move. On March 17, 1942, the 2nd embarked at Glasgow as 5th Division proceeded overseas. On May 6, it disembarked at Madagascar in the brief onslaught on the Vichy French that was under the overall direction of one of its old Commanding Officers, Lieutenant-General Sir William Platt, now G.O.C. East Africa after his successful Abyssinian campaign, but re-embarked a fortnight later for India. The 2nd, with 5th Division, was now well on its lengthy perambulations: in September, it entrained for northern Persia and wintered at Qum; the next February, it was outside Damascus for mountain warfare training in Syria, and that June, the call back to action came: the 2nd embarked at El Shatt, on the Suez Canal for the assault on Sicily.

The 2nd landed at Syracuse area on July 10 as 5th and 50th Divisions drove through feeble Italian defences to hardening German resistance along the coastal road to Catania. It was a brisk, strenuous campaign conducted in conditions of heat and dust over the arid Sicilian countryside, with another old foe of British soldiers in Sicily present, malaria. On July 18, as the Herman Goering Division held the River Simeto barring the approaches to Catania, the Wiltshires secured and held the vital only bridgehead enabling their brigade to cross the river. On

August 12, Sicily was clear and invasion preparations against Italy began.

On September 3, with the other invasion forces, the 2nd crossed the Straits of Messina that the old 62nd had known so well and landed north of Reggio. Winter set in with the 8th Army on the Sangro.

In the New Year, 5th Division left the 8th Army and joined 10th Corps in the 5th U.S. Army. It was soon faced with a major action, the crossing of the Garigliano river with two other divisions:

> 'The Garigliano was about 50 yards wide, deep and swift-flowing between high flood banks. The river plain was open, with ditches and drains. The Tufo feature (the objective of the 2nd Battalion's brigade) was rocky, covered with olive trees, broken by stone fences, with small buildings scattered along the slopes. A Recce Regiment was patrolling the river line.
>
> 'The German 74th Infantry Division had been in position for several weeks. Dugouts, trenches, wire and obstacles had been constructed along the forward slopes of the hills. Mines were plentiful and the line of the river was held by a strong screen.'

Thoroughly rehearsed, the 2nd moved up to the river line.

> 'On 17 January, the Battalion left its billets in TCVs, had a hot meal en route, after dark, and at 8 p.m., debussed and marched up to the FUP. The 56th Division, crossing on our left, woke up the enemy, and the Battalion was under slight fire while in the FUP waiting to cross. However, the crossing went according to plan and "B" and "D" Companies crossed and formed up on the open plain across the river. "A" Company was the bridgehead company, and "C" Company manned the assault boats.'

After some casualties from enemy mortar fire, the 2nd eventually pressed on when the other assault battalion of 13th Brigade was firm.

> 'At 2 a.m., on the 18th, the Battalion advanced across the plain in night attack formation, in half moonlight, well-

guided by the Intelligence Officer. . . . When it reached the Ausents stream, this was found to be a steep-sided and deep canal. Fortunately, a footbridge was found by "B" Company, by which the Battalion crossed and re-formed just as the barrage opened up; and as the sky lightened in the east, everyone pressed on to get up on the ridge before dawn broke.

'The barrage was so effective that little opposition was met until the leading companies reached Tufo village. "B" and "D" Companies found the enemy well-entrenched, and for some hours fought in the houses; and in the end, "B" Company held most of the village but the enemy occupied the western end. "D" Company soon reached their objective, on the saddle between the village and Point 156 ridge. "A" Company came up and took over the 102 spur from the Carrier Platoon, and "C" Company linked up "A" and "B" Companies.

'Many prisoners were collected within the Battalion position during the day; and just before dark, three German tanks attacked along the lateral road and railway, and shelled Battalion H.Q. at short range. The leading tank was hit by PIAT fire from the Carrier Platoon, and all three withdrew.'

But other hard days were to come, as the whole British position was precarious, with 56th Division still held in the river plain below. Heavy counter-attacks came in, and the final enemy thrust drove 'A' Company, with the Company commander and two platoon commanders killed and the strength down to thirty-one, off the key ridge. But the Cameronians retook the ground as the Germans thinned out to meet the threat in their rear of the Anzio landing.

'The outstanding feature of the battle was the determination and high morale of the Battalion. Its successful crossing and initial gains, due to the thorough preparation and the dash of the rifle companies, enabled the Division to get a strong foothold on its objective. In spite of short rations, frosty nights without blankets or greatcoats, and under heavy shelling, the Battalion hung on to the vital ground and thus enabled the Division to build up and consolidate on its objective while the other two divisions of 10 Corps were still struggling to establish themselves across the Garigliano.'

After holding defensive positions in this area, the 2nd landed in the beleaguered Anzio bridgehead with their Division and fought a close-locked battle there until the breakout two months later. Then, during a battalion attack on a strong enemy position on high ground outside Ardea, Sergeant M. A. W. Rogers, who had won the Military Medal in Sicily, now posthumously gained the Victoria Cross commanding the Carrier Platoon whilst both the two Carrier Platoon N.C.O.s were also killed in this splendid action.

Also, in this battle before the imminent fall of Rome, Major D. E. Ballantine and Captain R. A. S. Ward gained the M.C. When the Tiber was reached 5th Division was taken out of the line to rest and one month later, the 2nd Battalion, after sustaining 600 battle and sickness casualties in three months, was transported to the restful setting of Palestine and Syria.

The onset of 1944, meanwhile, in faraway India saw the long overdue approach of the 1st Battalion to battle. In late 1943, 200 short of establishment and badly weakened by the repatriation of some 270 long-service N.C.O.s and men, the Battalion began jungle training in Bengal. As one of the only two British battalions in 26th Indian Division, it was to be part of 15th Corps advance into the North Arakan. When 7th Indian Division was cut off by a prodigious Japanese flanking march, the Wiltshires, as part of 4th Indian Infantry Brigade, were ordered in early February to advance to their relief.

The 1st Battalion was dug in at North Bawli Keep when they were ordered to move by night to Goppe Bazaar over the hills, supported by some 500 mules. At Goppe Bazaar on February 13, the battalion became Advance Guard to press on to the relief of 7th Indian Division in the Sinzweya Box.

The tactical advance over the intervening miles of rude Burmese track flanked by jungle covered mountains was hard and laborious. The first night out, 'C' Company was attacked in strength but the enemy were driven off in a spirited counter-attack. Supply was by air and the Japanese Zeros came in on strafing runs, attracted by the bonfires lit to guide the air drops. By February 15 the Battalion had pushed on to Yinbounggyaung, and cleared an enemy position on a feature with the codename Wings, though not before the enemy fired the padi stubble and caused considerable stress to the helpless wounded.

The Battalion was now probing deeply into the Japanese defence system about the surrounded 7th Division, and February 22 saw a battalion attack on a dominating enemy position called Three Pimples.

> 'On 22 February, these Bunkers were attacked by "A" Company from the north and "D" Company from the east. . . . However, the Japs could not be displaced and the Battalion stood back to enable the medium artillery to soften it up. After further probing by "C" and "D" Companies, the position was captured at 11 a.m. 25 February, and the mountain ridge was then discovered to be a honeycomb of Japanese foxholes which were now full of enemy dead. The next day the Battalion passed through the old 7th Division H.Q. which still showed the havoc wrought by the Japanese attack. Broken tents, typewriters, wireless set, camp beds and decomposed corpses lay scattered everywhere.'

On March 1, the Wiltshires entered the 7th Division box and now dug in a defensive position at Tatmin Chaung to go into hard-earned reserve. In early April, reinforcements, though none of them were Wiltshires, arrived to revive the depleted ranks and were in time to participate in the battle for Pt. 551.

> 'The Divisional objective was now Pt. 551 which overlooked the Tunnels and commanded most of the road to Buthidaung. It was soon obvious that to reach this Jap stronghold it would be first necessary to attack and capture

two features known as Hump and Camel, which were also strongly held by the enemy. The Wiltshires were selected for this task, and then ordered to pass a Company through the gap and to capture two further features known as Barrel and Bullet. The Battalion was given support by a squadron of tanks (25th Dragoons) and a regiment of medium artillery. The C.O. decided to put in a setpiece attack on Hump and Camel, and at first light on 15 April, the battle started. During the night the artillery had kept up a constant pounding of the enemy position, finishing up with a twenty minute concentration. During this time, the tanks had moved up into position. The plan was for "B" and "D" Companies to attack the two positions simultaneously, for "C" Company to exploit the success and for "A" Company to be in reserve. At zero hour 09.00 hrs, the leading companies attacked their objectives. However, unfortunately, the dust from the artillery fire was so thick that the tanks could not see well enough to cover the advancing infantry. Both features were steep and covered with dense foliage, and the men had to pull themselves up clutching the bamboos with both hands. Fighting was intense and, despite the punishment the Japs had received from the artillery, they machine-gunned the advancing Wiltshires from both sides. However, within an hour, Camel was captured, although "D" Company had been driven back and suffered severe casualties. The Company Commander, did his utmost to penetrate the position but in doing so, he was himself shot through the legs. However, with Camel in their hands, the Wiltshires re-organized and in the afternoon, attacked again. "D" Company had received 40% casualties, but together with "B" Company, they attacked the enemy from two flanks and moved up on the Hump with the tanks firing point blank over their heads. This time they were successful, and when darkness fell, Hump was captured and the Japs were in full retreat. Despite mortar fire and counter-attacks, the Battalion hung on to their position, and the next night "A" Company passed through and with little opposition, captured Bullet Barrel. Fighting was now intense. The Japs constantly mortared and sniped all the company positions. It became impossible to move by day and mules had to bring up supplies by night. However, even so it was extremely difficult to administer the leading companies because casualties now became heavier, Captain Canning was killed and the C.O., Lieutenant-Colonel J. Houghton Brown was seriously wounded, the Japs kept up constant attacks with fighting patrols.'

The Wiltshires were now relieved in their forward positions by the 2/7th Rajputs and soon afterwards, pulled out of the Arakan as 26th Indian Division was redeployed to block the Japanese thrust on Imphal. The monsoon was spent in positions astride the road from Bawli Bazaar to Maungdaw but despite reinforcements, the Battalion was in bad shape physically, having suffered over 200 battle casualties and three times this number from the malarial scourge. In October, 1944, the Wiltshires were withdrawn to Rawalpindi, in the Punjab, to recuperate.

While the 1st Battalion was enmeshed in the jungles of Burma and the 2nd Battalion drew out of the Italian

Imperial War Museum

Carrier-borne troops of the Wiltshires push on during the advance, Normandy, July 1944.

campaign to rest, it was now the turn of the finely trained 4th and 5th Battalions to march into battle. As a follow-up formation to the invasion, their Division, 43rd (Wessex), landed in Normandy during June 24-26, 1944.

Soon, in early July, both battalions were embroiled in fierce close fighting in the villages and woods of the 'Odon Box'. The Wiltshires moved across the River Odon to join in the Division's first major battle, for Hill 112 and the vital ridge of high ground between the rivers Odon and Orne.

In the dust and heat of high summer, amid sunken lanes and cornfields, the divisional attack went in at first light on July 10 and encountered heavy artillery fire. An hour later, the 4th and 5th Battalions, with the 4th Somersets, had gained the crown of the hill, with Churchill tanks rumbling and clanking behind in support, but the German main positions were on the reverse slope and much fighting was still to come.

Despite the powerful enemy counter-attacks that developed that afternoon, the Wiltshires, with the 4th Somersets and 5th DCLI, hung on to the vital Hill 112. During that long hard day, they had sustained well over 200

Imperial War Museum

Wiltshiremen advancing south of Caumont, July 1944.

casualties. The whole division remained in this area for three active weeks, until the 4th and 5th Wiltshires, supported by 7th Royal Tanks, resumed the offensive to capture Maltot during July 22 and 23 after a savage battle that yielded 400 prisoners.

After a brief rest, 43rd Wessex re-entered the line at Caumont on August 1 and fighting through extensive minefields and considerable tank opposition in the hilly and wooded countryside, the West Country regiments were in position by August 5 to assault the dominating Mont Pincon. During the approach, the 4th and 5th Wiltshires each led the Division on separate axes, and with the misty dawn of the 5th, a day of stifling summer heat and exhausting battle developed for both battalions. Two miles south-west of Mont Pincon, the 4th waded the stream of the River Druance west of St. Jean Le Blanc but as they fought through the hot day on high ground across the stream, in the north the 5th ran into considerable entrenched enemy opposition about the stream west of La Varinière; one company got across but that night under great pressure, the company, whose commander, Major Thomas, one of the young officers at Bangalore in 1939, was killed in the withdrawal, fell back across the river. The 4th Battalion was also withdrawn back across the stream into reserve at Danvou.

During the afternoon of August 6, the battle was resumed, with the 4th Somersets and 4th Wiltshires, supported by the tanks of 13/18th Hussars, going for Mont Pincon again. Chester Wilmot, in his *The Struggle for Europe*, depicts the scene:

> 'The day was hot and sultry. The still air was laden with dust and the acrid smoke of explosives and the sickly stench of dead cattle bloated by the sun. Every moving vehicle stirred up more dust, and drew fire from the enemy and curses from the infantry, who lay in slit trenches awaiting word to move across the river and up the steep scarp. The valley was heavy with foreboding as it echoed the roll and crack of guns.

> 'Shortly before three o'clock, under cover of smoke, pioneers lifted the mines from the bridge on the road to La Variniere and then the barrage came down on the far bank. Soon the barrage started to creep up the hill and immediately the 5th Wiltshires began wading the stream. They were still struggling across when the Germans came to life with machine-guns and mortars. In a matter of minutes the two leading companies were crippled and disorganized.'

In the north, the Somersets suffered a similar fate in the stream west of La Roguerie, where mines prevented their tanks crossing. With the 5th Battalion, a dramatic prising-open of the enemy defence occurred.

> 'Seeing his men pinned down in the bed of the stream Lieutenant-Colonel Pearson deliberately strode down the road to the bridge, swinging his walking stick, with a red rose showing defiantly in his steel helmet. The men in the stream-bed were cheered by this fine gesture and pressed forward as Lieutenant-Colonel Pearson fell, shot through the heart by a German sniper.'

A corporal immediately shot dead the sniper as the 5th surged over the stream, inspired by the noble gallantry of their Commanding Officer. The tanks of their supporting 13/18th squadron then found a way over and together, the Wiltshires and the tanks fought through La Variniere, taking 100 prisoners, to the crossroads at the east end where in the early evening the Brigade Commander wisely ordered the hard-tried battalion to go firm. The 5th, now commanded by the Adjutant, was down to a fighting strength of five officers and seventy N.C.O.s and men.

The 4th Battalion, as Brigade reserve, was now directed through to exploit the hard won success of the 5th. About the same time another remarkable feat of arms occurred; finding an unguarded track, two troops of the 13/18th Hussars drove hard up it and though one slipped and overturned and another was hit, the others appeared brazenly at the summit. The C.O. of the 4th, then led his battalion, very tired and depleted after its battle at St. Jean the day before, its night withdrawal of seven miles into reserve and with casualties from shellfire *en route*, in the fading light audaciously up lanes and by-ways to the high ground where the lonely and brave British armour waited. So ended a hard and exhausting day in the campaign for Normandy, under a burning sun and in choking dust that parched tortured throats, that ended in sweet victory for the dog-tired Wiltshiremen.

By mid-August, the whole German front had collapsed, culminating in the slaughter of the Falaise Pocket. 43rd Division was now ordered to mount the first assault crossing of the Seine, and 129th Brigade was its chosen assault brigade to make the crossing on night August 25/26. The operation was extraordinary in that it was mounted some 120 miles from the river, with the two Wiltshire battalions carried in D.U.K.W.s for the last ninety miles. Again the two assault battalions were the 4th Somersets and the 5th

Imperial War Museum

The 5th Wiltshires cross the Seine, August 1944.

Wiltshires and in this most difficult of operations, an opposed river crossing, again the battle was bloody and confused.

'All went well till both boats hit the shallows. The wind had lifted the smoke screen and the enemy brought down murderous MG fire from positions on the far bank. One boat was pushed off from the shallows into deeper water where it sunk with the loss of all on board. The Platoon Sergeant in the other boat ordered the men to swim for it, but he was the only one to reach the far bank. He then swam back to the boat and with the help of the R.E. crew succeeded in bringing it back to the near bank.

'By now the remainder of the storm boats had been launched, but the smoke screen was no longer completely effective, and all came under heavy fire. Soon, all the boats except one had been sunk. The surviving one, commanded by an R.E. officer, suitably dressed in only a duffle coat and a pair of army socks, succeeded in making several journeys and ferrying over the remainder of "A" Company. By 20.30 hrs, "A" Company, now reduced in strength to 2 officers and about 40 O.R.s, had concentrated on the far bank and were moving on to their objective which was taken just before last light.'

With the storm boats gone, the problem now was to get the other companies across.

'At 22.30 hrs. the C.O. gave the order for "C" Company to cross in DUKWs. Four DUKWs took to the water and three bellied on the mud. The remaining one reached the far bank with one platoon who made their way to the Company RV.

'While "C" Company were sorting themselves out, "D" Company were ordered to cross in the remaining DUKW.'

Meanwhile, the depleted 'A' Company, as darkness fell, was beginning to meet enemy platoon-scale counter-attacks. As these were repelled, casualties mounted and ammunition began to run low. To aggravate the strain and confusion, both the radios of the Company Commander and his artillery F.O.O. had been damaged and the line party had been in a boat that had drifted downstream.

At 3 a.m., almost all the 5th were over the Seine, but now the gallant 'A' Company was overrun by an enemy company. However, 'B' Company managed to scramble into the exposed flank, taking under command seventeen survivors from 'A' Company, and at first light, the 5th Battalion, after a truly strenuous night of crisis and battle, were firm about their initial bridgehead.

At 10 a.m., August 26, the 4th Battalion were able to cross over the road bridge and by 6 p.m., another two battalions crossed over a folding boat bridge thrown across by the Royal Engineers under shellfire. August 27 saw desperate enemy counter-attacks that were repulsed and by that evening, the bridgehead was four and a half miles deep.

It was through this, the Vernon bridgehead that the whole of the British armoured units poured for the great drive to Brussels and Antwerp, achieving the liberation of northeast France and most of Belgium.

After a long overdue rest by the Seine for a fortnight, both the 4th and 5th followed up the triumphant advance of the British Army to an autumn and winter of hard, bitter fighting in the Nijmegen area, the Geilenkirchen offensive in November, the Roer Salient in mid-January, to severe fighting about Cleve and Goch, the breakthrough to the Rhine, and the drive through northern Germany to the ultimate investment and capture of Bremen in late April. It was at this time that the 2nd Battalion, concentrating with the 5th Division at Goch in April, crossed the Rhine at Zanten where the two Territorial battalions had fought a few weeks earlier and when hostilities ceased on May 8, reached Lubeck on the Baltic.

Again, in the great canvas of such a major war, it is impossible to depict with any justice the day-to-day heroism, endurance and suffering of all those Wiltshiremen who fought in battle.

The term 'Wiltshiremen' is used in the widest sense, as

Imperial War Museum

Men of the Wiltshires round up German prisoners, May 1945.

many of the wartime soldiers necessarily came from the whole nation. However, once inducted into the Wiltshire Regiment, they strove and fought against the maniac ambitions of Hitler and the grandiose dreams of the 'Greater Asia Co-Prosperity Sphere' as bravely and victoriously as their valiant predecessors in red or khaki had ever done.

Chapter 10

The Last Post

On a bitterly raw New Year's Day in the German winter of 1949, a significant event of post-war retrenchment occurred: the 1st and 2nd Battalions were 'amalgamated' at Krefeld before the Colonel of the Regiment, General Sir William Platt, G.C.B., G.B.E., D.S.O. The splendid citizen soldiers of the 4th, 5th and 7th Battalions had long returned to civil life after their yeoman wartime service.

In April, 1950, the newly constituted 1st Battalion left the 2nd Division and Rhine Army, to proceed to the Communist terrorist emergency in Malaya. However, though a large advance party trained in jungle warfare in Malaya, regrettably the Battalion never did relieve the Devons at Bentong, Pahang. At Colombo, the main body heard over the troopship's B.B.C. news that the 1st Wiltshires were diverted to Hong Kong as the Korean War had broken out. Three frustrating years were spent digging and holding defensive positions in the hills of the New Territories, with false starts to both Malaya and Korea, before the Battalion returned to its home county to the selected task of Demonstration Battalion at the School of Infantry, Warminster.

Active service came in January, 1956, when the Battalion moved to Cyprus. For three years, it patrolled, ambushed and searched for E.O.K.A. terrorists in the beautiful Kyrenia range district, with detached companies at Myrtou, later Lapithos, Kyrenia Castle and Ayios Amvrosios. The climax of these prolonged internal security operations came in November, 1958, when Matsis, Grivas' second-in-

Wiltshire Gazette

Post war: the Best Recruit of a National Service squad at the Depot, Le Marchant Barracks, Devizes, 1955, receiving his award from Colonel J. Houghton Brown, D.S.O., T.D. With him is Major The Duke of Somerset, Depot Commander, Major F. R. E. Turner, OC Training Company, and Captain T. A. Gibson, Adjutant.

command, was trapped and killed at Kato Dhikomo, three wanted men and some ninety E.O.K.A. smaller fry with large caches of arms and explosives taken in a most successful large-scale operation controlled by the 1st Wiltshires. This broke the E.O.K.A. organization in Western and Central Cyprus and significantly, Grivas sued for a truce soon afterwards. When the Battalion moved out of Kyrenia to embark at Limassol for England in late February, 1959, peace was in the air; gone was an era of waterpipe bombs thrown in the blackness of the night, the assassin's bullet and the road bend ambush, and the electrically detonated mine.

But the Wiltshires were landing on their native soil for a sad duty. Further retraction of the Army had now decreed

The Colonel-in-Chief, H.R.H. Prince Philip, Duke of Edinburgh, with officers of his Regiment at Buckingham Palace: Lt.-Col. A.E.G. Brain T.D., Major A. D. Parsons, M.C., Major S. F. Browne, Lt.-Col. H. M. A. Hunter, D.S.O., M.B.E., and the Colonel of the Regiment, Major-General B. A. Coad, C.B., C.B.E., D.S.O.

that the Wiltshire Regiment was to amalgamate with a sister County regiment, the Royal Berkshire Regiment, in June. The funereal ritual of the winding-up of 200 years of history began. On May 2 the Colours of the Regiment were trooped for the last time at Albany Barracks, Isle of Wight, before the Colonel of the Regiment, Major-General B. A. Coad, C.B., C.B.E., D.S.O., and two days later, the Colours were laid up in Salisbury Cathedral. To the strains of 'Auld Robin Grey', from the old 99th, the Colours were slow-marched to the Sanctuary step, where the Commanding Officer delivered them into the hands of the Bishop.

On June 9, 1959, amalgamation with the Royal Berkshire

A battalion operation against Eoka terrorists in the Kyrenia mountains, Cyprus, in September, 1958. Watching from a forward observation post with the Commanding officer, Lt.-Col. G. F. Woolnough, M.C., are Lt. H. D. Canning, Signal Officer, CSM D. C. Mortimer, CSM B. Coy, Major F. R. E. Turner, Second-in-Command, and Major F. J. J. Worth, OC A Coy.

Regiment occurred at a ceremonial parade on the Isle of Wight. The new regiment, The Duke of Edinburgh's Royal Regiment (Berkshire and Wiltshire) paraded for the first time before its Colonel-in-Chief, Prince Philip, Duke of Edinburgh. The Regimental Depot at Devizes lingered on until March, 1960.

But it was to be the 4th Battalion, the Territorials and men of Wiltshire, who struck the last chord of history for the Wiltshire Regiment. On March 12, 1967, on the demise of the old Territorial Army, the Colours of the 4th Battalion were also laid up in Salisbury Cathedral after a farewell parade of the Battalion in the Market Place. The Honorary Colonel, Brigadier E. L. Luce, D.S.O., O.B.E., T.D., D.L., who had commanded in N.W. Europe, read the Story of the Colours from the Nave Lectern to a packed congregation of Old Comrades.

The Wiltshire Regiment had finally returned to the people of Wiltshire.

Regimental March

The March *The Wiltshire* is based on an old Wiltshire tune
The Vly

1st Verse

T'were on a jolly summer's day, the twenty fust of May,
John Scruggins took his turmut hoe, with this he trudged away,
Now some volkes they loike haymakin', and some they vancies mowin'
But of all the jobs as Oi loike best, gie Oi the turmut 'oein.

Chorus The Vly, the Vly —
The Vly be on the turmut,
Tis all me eye,
For Oi to try,
To keep Vlys off them turmuts.

2nd Verse

The fust place as Oi went to wurk; it were wi' Varmer Gower,
Who vowed and swore as how Oi wer'—a virst class turmut oer;'

The second place Oi went to wurk, they paid Oi by the job,
If Oi'd a knowed a little more, Oi'd sooner bin in quod.

3rd Verse

The last place as I went to wurk, they zent ver Oi a mowin',
Oi sent word back, Oi'd sooner get the zack, than gie up turmut 'oein',
Now all you jolly varmer chaps, what bides at 'ome zo warm',
Oi'll now conclude my ditty wie a wishing you no 'arm.

1st Battalion The Wiltshire Regiment

(Duke of Edinburgh's)

(formerly 62nd [Wiltshire] Regiment of Foot)

1756	Raised in Exeter as 2/4th Regiment of Foot.
1758	Re-numbered 62nd Regiment.
1758	Siege of Louisbourg.
1759	Battle of Quebec.
1760	Siege of Carrickfergus Castle.
1760—63	Ireland.
1763—69	Dominica.
1763—75	Ireland.
1776—81	America (American War of Independence).
1781—84	Lincoln and Newark, Windsor, Trowbridge.
1784—90	Ireland.
1791—97	Jamaica and San Domingo (Maroon War 1796—97).
1797—1800	Exeter, Poole. 2/62nd Regiment raised.

1/62nd

1800—06	Ireland.
1806—14	Egypt, Sicily, Italy (Napoleonic Wars). Grenadier Company Spain 1813—14.
1814	Invasion of Maine (American War of 1812).
1814—23	Nova Scotia.

2/62nd

1800—02	Ireland.
1802	Disbanded.
1804	Reformed Devizes.
1804—09	England.

1809—13	Jersey.
1813	Ireland.
1813—14	Invasion of Southern France (Napoleonic Wars).
1814	Ireland.
1815—16	France.
1816	Disbanded in Ireland.

62nd

1823—30	Ireland.
1830—34	Southern India.
1834—40	Burma.
1840—47	India (First Sikh War 1845—46).
1847—49	Winchester, N.W. England.
1849—54	Ireland.
1854	Malta.
1864—56	Crimea (Crimean War).
1856—64	Nova Scotia, Canada.
1864—69	Ireland, Aldershot.
1869—80	India.
1880—81	Aden.
1881—95	Woolwich, Aldershot, Jersey, Ireland.
1895—1909	India.
1909—13	South Africa.
1913—14	Tidworth.
1914—19	France (The Great War).
1920—22	Ireland.
1922—29	Tidworth, Plymouth.
1930—32	Egypt.
1932	Shanghai.
1932—36	Malaya.
1936—43	India.
1943—44	Burma (North Arakan).
1944—47	India.
1947—48	Corsham and Monkton Farleigh.
1949—50	Krefeld.
1950—53	Hong Kong.

1953—55	Warminster.
1956—58	Cyprus.
1959	Amalgamation with the Royal Berkshire Regiment (Isle of Wight).

2nd Battalion The Wiltshire Regiment

(Duke of Edinburgh's)

(formerly 99th [Lanarkshire] Regiment of Foot)

1824	Raised in Glasgow.
1825—37	Mauritius.
1837—41	Ireland.
1841—56	Australia, New Zealand (First Maori War 1846—47).
1856—58	Ireland, Aldershot.
1858—59	India.
1860—65	China, Hong Kong (China War of 1860).
1865—69	South Africa.
1869—79	Scotland, Ireland.
1879—80	South Africa (Zulu War).
1880—81	Bermuda.
1881	South Africa.
1882—95	India, Burma.
1895—99	Portsmouth.
1900—03	South Africa (Second Boer War).
1903—12	Pembroke Dock, Ireland, Gosport.
1913—14	Gibraltar.
1914	England.
1914—19	France (The Great War).
1919—22	Hong Kong, Peking.
1922—29	India.
1929	Shanghai.
1930—36	Plymouth, Aldershot.
1936—37	Palestine.
1938—39	Catterick.
1939—45	France, England, Madagascar, India, Persia, Syria, Palestine, Sicily, Italy, N.W. Europe.
1945—48	Hannover, Krefeld.
1949	Amalgamation with 1st Battalion, Krefeld.